"Brian Keating shares a view from the jagged frontiers of scientific exploration, offering fresh insights into the passions, ambitions, and competition that drive many researchers today. A fascinating journey."
—David Kaiser, professor of physics and the history of science, MIT, and author of *How the Hippies Saved Physics*

"According to Brian Keating, the Oscar and Nobel science prizes have a lot in common. In *Losing the Nobel Prize*, he weaves together the Nobel Prize institution, his personal life, and his own involvement in modern cosmology into a multifaceted and highly readable story. Providing a vivid picture of the adventurous and competitive world of cosmological research, he also suggests radical reforms to the venerable but perhaps outdated Stockholm institution."
—Helge Kragh, emeritus professor, Aarhus University and author of *Cosmology & Controversy*

"Cosmologists had thought that they had glimpsed a distant image of the first moments of the universe. Instead, this image turned out to be a 'smudge on the window': galactic dust once again bedeviling cosmologists. Keating conveys this exciting search through a personal tale of the ups and downs of cutting edge science."
—David Spergel, professor, Princeton University, and co-winner of the 2018 Breakthrough Prize in Fundamental Physics

"A fascinating blend of personal history and an honest behind-the-scenes look at high-stakes science. Brian Keating was at the origin of what appeared to be one of the most exciting discoveries in modern cosmology. His vivid storytelling brings humanity's search for the origin of the Universe to life."
—Jay Pasachoff, author of *Peterson Field Guide to the Stars and Planets*

"I loved this well-told tale of science, passion, and the pursuit—literally to the ends of the earth—of life's purest questions. Brian Keating weaves together a must-read drama of big dreams, awe-inspiring technology, and a belief in the power of science to solve any puzzle. He is thoroughly modern and forward facing, questioning the veneration of the Nobel Prize, and making the case with his heartfelt story that the real prize is in the science itself."
—Julian Guthrie, author of *How to Make a Spaceship*

LOSING THE NOBEL PRIZE

LOSING
THE
NOBEL PRIZE

A Story of Cosmology,
Ambition,
and the Perils of
Science's Highest Honor

~

BRIAN KEATING

W. W. Norton & Company
Independent Publishers Since 1923
New York London

For information about permission to reproduce selections from
this book, write to Permissions, W. W. Norton & Company, Inc.,
500 Fifth Avenue, New York, NY 10110

For information about special discounts for bulk purchases,
please contact W. W. Norton Special Sales at
specialsales@wwnorton.com or 800-233-4830

Manufacturing by LSC Communications, Harrisonburg
Book design by Lovedog Studio
Production manager: Anna Oler

Library of Congress Cataloging-in-Publication Data

Names: Keating, Brian (Brian Gregory), author.
Title: Losing the Nobel Prize : a story of cosmology, ambition, and the
    perils of science's highest honor / Brian Keating.
Description: New York : W.W. Norton & Company, [2018] | Includes
    bibliographical references and index.
Identifiers: LCCN 2017057199 | ISBN 9781324000914 (hardcover)
Subjects: LCSH: Big bang theory. | Astronomy—Awards | Cosmology. |
    Science—Methodology. | Nobel Prizes.
Classification: LCC QB991.B54 K43 2018 | DDC 523.1/8079—dc23
LC record available at https://lccn.loc.gov/2017057199

W. W. Norton & Company, Inc.
500 Fifth Avenue, New York, N.Y. 10110
www.wwnorton.com

W. W. Norton & Company Ltd.
15 Carlisle Street, London W1D 3BS

1 2 3 4 5 6 7 8 9 0

To my mother, Barbara, who gave me life,
and to Sarah, who gives me life.

CONTENTS

A NOBLE WILL

Each year, on December tenth, thousands of worshippers convene in Scandinavia to commemorate the passing of an arms dealer known as the merchant of death. The eschatological ritual features all the rites and incantations befitting a pharaoh's funeral. Haunting dirges play as the worshippers, bedecked in mandatory regalia, mourn the merchant. He is eerily present; his visage looms over the congregants as they feast on exotic game, surrounded by fresh-cut flowers imported from the merchant's mausoleum. The event culminates with the presentation of gilded, graven images bearing his likeness.

This ritual is the annual Nobel Prize award ceremony, but you'd be forgiven for thinking it was an occult sacrament. The focus on death might be surprising. But, in truth, the Nobel Prize was born in death. The death of Alfred, inventor of dynamite, birthed the prize that redeemed the Nobel name, achieving the ultimate in posthumous public relations.

The Nobel Prize is not merely science's most esteemed accolade, but the *world's* most prestigious award. Its goal: to recognize humanity's crowning achievements in the sciences, free from ideological agendas or fashion, as well as to reward poets and peacemakers who enrich mankind. When it's "done right," it is a truly lustrous meritocratic reward system. Yet, in this book, I explore the case *against* the Nobel Prize in Physics, arguing that its days may be numbered unless it is radically reformed.

In a society that seemingly has more awards ceremonies than talented celebrities, you might wonder what's wrong with a single award celebrating the most honorable of goals, the betterment of mankind via science? I would answer that, like a coin, the Nobel Prize has three sides. Its obverse, or positive side, corresponds to the respect and admiration it brings to science and scientists. Its reverse, or negative side, reflects how it punishes collaboration and causes ferocious competition for scarce resources. Lastly, the gilded medallion's unstable rim evokes the prize's uncertain future in our modern scientific era. Many young scientists now ask: Is the Nobel Prize a fair coin?

This book is not a polemic, nor is it a scorched-earth attack on the prize. Instead, I offer a unique, insider's glimpse into the power of the prize to refract reality, as it did for me, an astronomer who was once seemingly about to read the very prologue of the cosmos. For thirty years I had been mesmerized by the prize's golden promise; eventually, I designed a Nobel-worthy experiment, only to have the medal slip from my grasp. The end of that dream removed a veil that had been covering my soul as a scientist. I began to see the prize not as a divine anointing but as a human creation, with some all-too-human frailties. Of course, for a time, I felt disappointment, anger, and maybe a bit of bitterness—but those are not the animating spirits of this book.

Losing the Nobel Prize was born from a genuine desire that the Nobel Prize in Physics achieve the great purpose originally envisioned for it: to be a beacon of excellence for *all* physicists. The story of the Nobel Prize, and my own narrative, are human stories—tales of idols and ideologues, pride and prestige, cunning and deception, shame and redemption. But most of all, this book is about the creative motivating passion that impels all scientists to venture into the abyss, one step at a time.

STOCKHOLM'S SKY WAS sublime. It was late summer, and an inky blue canopy beckoned him outdoors. Gentle winds caressed the windows of the laboratory, teasing him like a lover's whisper: *come play with me*. But Emil Nobel *was* playing. Though Heleneborg, the stone-walled laboratory of Nobel & Sons, had every bit the appearance of a jail, Emil must have felt free within its confines; the space emboldened him to try new tricks. Perhaps this time he could dupe Mother Nature into revealing just one secret more than she'd intended. Emil, the last and favorite of Immanuel Nobel's seven children, three of whom had died in childhood, found his home in the laboratory.

Each successful experiment delighted him: a magic trick, one whose prestige only he could reveal. Out of sight of his oppressively doting father and his older brothers, Emil could play. Here he, and perhaps he alone, could save Nobel & Sons, a family business built on death itself, the once preeminent manufacturer of highly explosive war supplies.

But Nobel & Sons itself was dying. The Russians had severely cut back their purchases of Nobel's boat-busting naval mines, the company's flagship product. Ludvig Nobel—the oldest son, to whom Immanuel, the patriarch, had transferred the operation of the business—tried to pivot. He bet the family fortune, such as it was, on drilling for oil. In the mid-nineteenth century, oil was typically harvested from whales and used mainly in lamps; automobiles were still five decades in the future.

Emil was fascinated by explosives. Though they demanded respect, he reasoned, what could be simpler than nitroglycerin blasting oil? Some nitric acid from the stock shelves, a bit of Skånian beef tallow for the glycerin. All he had to do was to stabilize it and it could be transported—even to America, where a westward railway expansion was underway. For once, a Nobel invention could be used for *con*struction. But how to tame it? Perhaps if the volatile oil was cooled, Emil hypothesized, it could be solidified into a stable mass, like butter. Why not? After all, the fatty glycerin had once girded the bellies of milk cows. Then it could be transformed into "Blasting Butter," friendly-sounding, safe to handle, yet potent when employed by a properly trained magician. Yes, Heleneborg was *his* playground—let his older brothers seek fortune in whale blubber substitutes elsewhere.

~

THE PRESSURE WAVE reverberated through the stone-walled laboratory at thirty times the speed of sound; Emil literally couldn't hear it coming. He'd stolen the Sun's heat and brought it down to Earth. As with Prometheus, it didn't end well.

~

THE EXPLOSION TOOK Emil's life along with the lives of four coworkers. Alfred was wounded too, though not severely. Neither he nor his father fully recovered from the devastating trauma. Immanuel suffered a debilitating stroke soon after Emil's accident and would be united with his son again in death: dying eight years to the day after Emil.

Nitroglycerin remains a chimerical compound, extremely sensitive to sudden jolts and electrical discharge. In fact, physical jolts, such as an impact with a hammer—or, as likely happened in Emil's case, being dropped—are a sure means of detonating it. More insidiously, it becomes less stable as it ages, making it extremely hazardous to transport. (None of these well-known hazards deterred the author, at fifteen, from attempting to win the Westinghouse Science Competition by synthesizing nitroglycerin from household chemicals available at the local A&P grocery store. Luckily, those attempts fizzled without incident.)

Of all Alfred Nobel's 355 patents, the most famous, and profitable, was the combination of nitroglycerin with absorbers and inert stabilizers. This invention allowed the explosive to be detonated in a controlled fashion, using electric or pyrotechnic charges (such as the fuses Bugs Bunny's mortal enemy Elmer J. Fudd futilely employs). The ideal stabilizer turned out to be a common antacid. The formerly volatile liquid could then be solidified by absorption into a chalk-like powder called diatomaceous earth (another pharmaceutical, still found in some leading toothpaste brands today). In an ironic twist, the deadly explosive was made entirely of edible ingredients (fat, antacid, and nitroglycerin). Alfred, like many who suffer from angina, would himself be prescribed nitroglycerin. This tickled him, as he wrote in a letter to his friend Ragnar Sohlman: "Isn't it the irony of fate that I have been prescribed nitro-glycerin, to be taken internally! They call it Trinitrin, so as not to scare the chemist and the public." Alfred recognized that putting the proper spin on your discovery was crucial; he originally christened his new explosive "Nobel's Safety Powder."

In 1867, a mere three years after Emil's death, Alfred received a patent for what he later called dynamite (from the Greek for "power rock"). Despite the fortune it brought him, guilt must have gnawed at him: if only he had been visited by Genius a few years earlier, his baby brother's life, his father's sanity, Nobel & Sons' fortune, all would have been spared. Dynamite made Alfred one of the richest men in the world, but death continued to stalk the sons of Immanuel Nobel.

KILLING NOBEL

Much lore surrounds Alfred Nobel and the prizes that bear his name. Some of these tales explain why there is no Nobel Prize for mathematics, or recount indecorous narratives about Alfred's wife (even though he never married). Of all such tales, only one, the Prize's origin story, as outlandish as it sounds, may actually be true. The story alleges that while in Paris in 1888, Alfred read an obituary purporting to be his own: *Le marchand de la mort est mort*

("The merchant of death is dead"). Alfred was startled. It described him—the inventor of dynamite—as *the* single person who had enabled men to kill one another faster and more efficiently than anyone in history. Alfred was, of course, very much alive in 1888; it was his older brother Ludvig who had passed away recently in Cannes. Alfred and Ludvig, like Alfred and Immanuel, had had a falling-out, but thankfully the brothers had reached a rapprochement a year earlier. Ludvig's obituary shocked Alfred, causing him to reconsider his own mortality.

Alfred loved France, but the feeling wasn't mutual. He offered to sell his explosive technology to the French government, but they declined: they had heard he'd offered it to the Italians, their adversary at the time. The French saw Alfred as an enemy of the state, and he was forced to flee. In retrospect, the mistaken obituary must have seemed like the Parisians' wishful thinking.

Alfred settled in San Remo, Italy, only returning to Paris once, in 1895, where he clandestinely wrote out his will by hand. A year later, he died from complications of angina pectoris. As an introverted bachelor with few friends, he kept his final testament secret until it was read after his death. The bombshell within was worthy of an explosives inventor:

> *The whole of my remaining realizable estate shall be dealt with in the following way: the capital, invested in safe securities by my executors, shall constitute a fund, the interest on which shall be annually distributed in the form of prizes to those who, during the preceding year, shall have conferred the greatest benefit to mankind. The said interest shall be divided into five equal parts, which shall be apportioned as follows: one part to the person who shall have made the most important discovery or invention within the field of physics. . . .*[1]

AN OFFER I COULDN'T REFUSE

On October 13, 2015, a week after the announcement of the 2015 Nobel Prize winners, I arrived at my office at UC San Diego's Center for Astrophysics and Space Sciences to find an intriguing communiqué from the Royal Swedish Academy of Sciences. "That's odd," I joked to one of my graduate students, "they should have contacted me last week if it were the news I deserved." Concealed within was a valuable document, one that would ultimately take me on a journey of introspection and liberation.*

Professor Brian G. Keating

On behalf of the Royal Swedish Academy of Sciences we, as members of the Nobel Committee for Physics, have the honour of inviting you to submit a proposal for the award of

The Nobel Prize in Physics for 2016

Stockholm, September 2015
Anne L'Huillier
Chairman

At first, it seemed like a huge honor. Soon, however, it became a heavy burden. A year before I received the letter, I had been rele-

---

* The invitation letter said that I could nominate winners for a specific invention or discovery, but noticeably absent was the requirement that my nominee's discovery had conferred the "greatest benefit to mankind," as Alfred's will required. The transgressions/violations of his will didn't end there. The letter said that *multiple* people could win the prize, and concluded with an allowance to nominate discoveries made long ago, not made merely in "the preceding year" as Alfred's will required—as long as the significance of the finding had only recently been recognized.

gated to a footnote in the public announcement of a discovery that seemed to be a shoo-in for the Nobel Prize, despite having been the designer of the experiment. If I accepted the invitation to submit nominations for the prize, wasn't I tacitly propping up the very system I, and many others, felt needed fundamental transformation? Wouldn't it make me a sellout? My stomach knotted as a paralyzing bout of hypocrisy washed over me. Ethical dilemmas aren't common occurrences for practicing cosmologists.

My invitation to nominate came exactly fifteen years after I received my PhD. A typical scientist's career lasts about thirty years. Thus, I was officially at midlife, careerwise: as good a time as any to have a crisis. I was reminded of a speech John F. Kennedy gave in 1959, when he said, "When written in Chinese, the word 'crisis' is composed of two characters—one represents danger and one represents opportunity."[2] My invitation to nominate *was* an opportunity, a chance to help reform the Nobel Prize, to improve it, so that it could continue to be worthy of such unadulterated respect. I secretly hoped the effect of the other character—representing danger to my career, presumably—would be minimal.

There were instructions included in the envelope. The first told me not to speak about my invitation to be a nominator. So, as you read this, odds are that I have submitted my last Nobel nomination.[3]

"You may not nominate yourself," the instructions continued, helpfully reducing the list of potential nominees by one. Even I wasn't vain enough to nominate myself. I decided to begin my task, as any academic would, with a close reading of a primary source: Alfred Nobel's will. Why not give the dead man his due? I barely got through its first sentence—the stipulation that prizes go to discoveries made during the preceding year—before I suspected he might be rolling over in his grave.

The preceding year? What *physics* discovery or invention could have conferred benefit to all of mankind in 2015? How can one even determine the degree of benevolence of a physics discovery? Is nuclear fission—which, like dynamite, can both create and destroy—a net benefit to humanity?[4] The "benefit" condition stemmed from Alfred's *fin-de-siècle* dreams of a better world, one that arcs toward peace and is driven by the altruistic work of scientists. I wondered if anything in my field, astrophysics, had *ever* conferred as great a benefit on humanity in the way that the first Nobel Prize in Physics did.

~

ON NOVEMBER 8, 1895, Wilhelm Röntgen accidentally won the first Nobel Prize in Physics. Toiling away in his lab, he discovered a mysterious form of radiation coursing through his Vienna laboratory. When photographic plates are exposed to cathode rays—beams of electrons—they get developed, even when opaque aluminum foil completely covers the plates. Röntgen's wife, Anna Bertha, became the first hand-bone model in history when she put her hand between the rays and a piece of film. When she saw her skeleton's digits on the exposed film she cried out, "I have seen my death!" Fortunately, Anna went on to live for several more decades and Röntgen's rays, which you will recognize as X-rays, have saved or improved the lives of billions, not taken them.

The speed of application of the discovery—from physics lab to physician's office within a year—was unprecedented, and it was celebrated around the world. Alfred Nobel wrote his will just weeks after Röntgen's serendipitous invention. Though the prize was awarded to him six years later, Röntgen would become the paradigm for future Nobel Prize winners—a lone physicist whose discovery *immediately* bettered mankind. Quick, beneficial, and clean, just the way Alfred wanted it.

One hundred and fourteen years later, my invitation to nominate candidates for the 2015 Nobel Prize in Physics made it clear that the Nobel Committee wasn't adhering to Alfred's stipulation on timing. The dispensation allowing discoveries made long ago was a workaround that seemed to be the committee's way of recognizing the messiness of the modern scientific enterprise, where years—plural, not singular—are needed to elevate discoveries into the scientific canon. Only a handful of Nobel Prizes have been awarded the year after the prizewinning discovery or invention was made; some have been awarded for discoveries made nearly a half century earlier. How long, I wondered, had Alfred's requirement been treated as a mere suggestion?

This is the first of many departures from Alfred's will. Ultimately, it is a modification I agree with. Science takes time; decade-long experimental searches are common. Decades more are sometimes required to vet, and to replicate, the discoveries. This is time well spent; it confirms that the findings will stand the test of time, and prevents the rush to judgment, pro or con, that often accompanies the latest "scientific breakthroughs."

But even this stipulation comes with a dark side. As we will see in chapter 5, the decades-long prize process often outstrips the average human lifespan, with several prospective laureates dying just before their accomplishments became "fully appreciated," by the Nobel Prize committee at least.

Other modifications to Alfred's will are far more insidious. Not only might he not approve but he might not recognize what has become of his final testament. These departures have, I believe, distorted his altruistic vision for world-bettering scientific discoveries and, worst of all, affected the careers of physicists, especially younger ones.

Like the other five Nobel Prizes, the physics prize is encumbered by arbitrary strictures and concealed behind a secretive process. And while the discoveries physicists make in the basic sciences are usually less controversial than those, say, in economics or medical science—fields laden with ethical implications—the physics prize suffers from systemic biases, ailments that, sadly, are self-inflicted.

In its early years the physics prize was plagued by anti-Semitism, with Hitler's Chief of Aryan Physics, Nobel laureate Philipp Lenard, personally leading a campaign to prevent Einstein himself from winning one. Thankfully, that shameful chapter has long been abolished (and Lenard would surely be distressed to learn of the large number of Jewish laureates!). However, there have been only two female physics laureates, despite the much greater number of deserving candidates in over one hundred years. No woman has won the Nobel Prize in Physics since 1963.

The other five Nobel Prizes have also seen significant controversy over the years. The chemistry prize awarded to Fritz Haber in 1918 was criticized because Haber had used his discoveries to make chemical weapons.[5] The Nobel Prize in Physiology or Medicine awarded to Antonio Moniz "for discovery of the therapeutic value of the lobotomy" in the treatment of certain forms of mental illness led to the popularization of the technique, without regard to its questionable ethical implications. And the Nobel Prize in Economics, a prize Alfred did not include in his will, no longer exists; it is now known as the Sveriges Riksbank Prize in Economic Sciences in Memory of Alfred Nobel. Sure, it's catchy. But what would prompt such a radical reformation of the only social science to be associated with the Nobel name?

Even the prize that was closest to Alfred's heart, the peace prize, has seen much controversy, from its presentation in 1973 to principal actors behind the Vietnam War to its award in 1994 to Israeli and Palestinian leaders who can hardly be said to have "reduced the standing armies of the world," as Alfred's will specified. Indeed, three past laureates signed onto a lawsuit against the Nobel Foundation, charging that the award of the 2012 peace prize to the European Union, which does not "realize Nobel's demilitarized global peace order," violated the conditions of Alfred Nobel's will.[6] Heck, even the award of the 2016 Nobel Prize in Literature to a popular musician, Bob Dylan, caused an outcry.[7]

There is much to learn from how institutions fail to live up to their potentials, and the Nobel Prize is no exception. The tribulations faced by the five other prizes offer teachable moments for the physics prize—which is, I fear, in danger of being tarnished. Luckily, there is time to reform it before it is too late.

In this spirit, three chapters in this book (chapters 5, 10, and 13, set off with gray borders for convenient identification) diagnose the Nobel Prize's detrimental effects on science's three most essential elements: credit, cash, and collaboration. These "broken lenses," as I call them, distort the way science is perceived, especially by young scientists. Finally, in chapter 16, "Restoring Alfred's Vision," I offer suggestions for reforming humanity's superlative award, not merely by improving its "optics" but by offering substantive corrections so that it can best support the way science is done today. Together, these four chapters are an anatomy of the Nobel selection process as seen by an insider (though odds are that, by the time you read this, I will be an outsider).

At first, the Nobel Prize's liabilities didn't faze me. In fact, I was blissfully unaware of them for decades, during my formative years as a young scientist. Honestly, even had I known about them, I wouldn't have been deterred. Like many, I was mesmerized by the prize's lustrous allure. What net worth is to an entrepreneur, what an Oscar statuette is to an actor, what an Olympic gold medal is to an athlete—that is what the Nobel Prize means to many scientists. Its laureates are among the elite intellects of human history, so much so that my children's rooms are adorned not with posters of sports heroes but with images of physics laureates, from Maria Goeppert-Mayer to Albert Einstein.[8] I have enough books written by or about laureates—Richard Feynman, Steven Weinberg, Frank Wilczek—to fill a small library. I also confess to having read a few books with titles like *How to Win a Nobel Prize*, tomes I later learned were about as helpful as those that offer strategies to win the lottery.[9] And after-market Nobel Prize medallions are so treasured that if you want to buy one of your own, you'd better win the lottery; recently auctioned specimens have brought in between $400,000 and $4.75 million.[10]

Since the first prizes were awarded in 1901, the Nobel Foundation has given out more than a billion dollars. Of course, physicists don't seek the prize for the $1 million-plus purse, nor for the 18-karat gold medallion, nor for the all-you-can-eat reindeer dinner with the King of Sweden. No, what many seek is more valuable than all these perks put together: an eternal tribute. And what could be more noble than pursuing immortality through the betterment of all of mankind?

In *The Denial of Death*, Ernest Becker's opus on materialism, mortality, and meaning, time is the antagonist. From time immemorial, kings and pharaohs, princes and presidents, have erected shrines commemorating their own brief existence. According to Becker, we all have a compulsion for some aspect of our brief tenure on the planet to endure long after we have left it. Becker spoke of it as a *causa sui*, an animating impulse: a cause worthy of dedicating one's life to, even if only symbolically, in order to combat life's ultimate meaninglessness. Immortality comes at a price, however. Pyramids don't come cheap. Nor does the Nobel Prize.

A N D   N O W   I invite you on a journey into eternity, one that will take us to the ends of the Earth to glimpse the beginnings of time. It is a quest so obviously Nobel-worthy I knew it would change my life forever. What I didn't know, what none of us could know, was how this extraordinary experiment would change cosmology, and science itself, forever.

LOSING THE NOBEL PRIZE

Chapter 1

READING THE
COSMIC PROLOGUE

*"You may speculate from the day that days were created,
but you may not speculate on what was before that."*

TALMUD, TRACTATE HAGIGAH 11B, 450 CE

SOME SAY TIME BEGAN WHEN THE UNIVERSE did, in what is commonly called the Big Bang. Others say time is a continuum, without any beginning whatsoever. Still others favor cosmological theories that posit not one single "bang" but an infinite number of them. Fortunately for my theorist colleagues, their grant proposals aren't reviewed by fifth-century Talmudic scholars. Yet even these ancient sages have modern-day counterparts, cosmologists such as Stephen Hawking, who called the question of what preceded the Big Bang as nonsensical as asking, "What's north of the North Pole?"

Why are alternatives to the Big Bang, models which go by names such as the Steady State or bouncing or cyclic theories, so unpalatable to the brightest minds of cosmology? While they are certainly less familiar than the Big Bang theory, noticeably lacking sitcoms named after them, they have appealed to scientific luminaries throughout history, from Aristotle to Albert Einstein to modern-day cosmologists like Roger Penrose. Models without a Big Bang have come in and out of fashion more often than wide ties on Wall Street.

Some scientists, including Nobel laureate Steven Weinberg, say these alternatives offer an additional attraction to many secular scientists (and lay observers alike) because they "nicely avoid the problem of Genesis." If there was no Big Bang, you don't need a Big Banger.

Most intriguing of all, these alternative models handily answer the question of what preceded the Big Bang: it could be a crunch or a bounce, the violent death throes of a previous epoch that gave birth to our current cosmos. But these are all still just theories. What has always tantalized me is another question: could the tools of modern cosmology, computers and telescopes, human brains and ultrasensitive detectors—real observations and real data—ever tell us whether or not time itself began?

To go back to the beginning, if there was *a* beginning, means testing the dominant theory of cosmogenesis, the model known as inflation. Inflation, first proposed in the early 1980s, was a bandage applied to treat the seemingly grave wounds cosmologists had found in the Big Bang model as originally conceived, flaws that I will soon explain. To call inflation bold is an understatement; it implied that our universe began by expanding at the incomprehensible speed of light . . . or even faster! Luckily, the bandage of inflation was only needed for an astonishingly minuscule fraction of a second. In that most microscopic flash of time, the very die of the cosmos was cast. All that was and ever would be, on a cosmic scale at least— vast assemblies of galaxies, and the geometry of the space between them—was forged.

For more than thirty years, inflation remained frustratingly unproven. Some said it couldn't be proven. But everyone agreed on one thing: if cosmologists could detect a unique pattern in the cosmos's earliest light, light known as the cosmic microwave background (CMB), a ticket to Stockholm was inevitable.

Suddenly, in March 2014, humanity's vision of the cosmos was shaken. The team of which I had been a founding member had answered the eternal question in the affirmative: time did have a single beginning. We had proof. It was an amazing time indeed.

March 17, 2014

FOR WEEKS I had known it was coming. Our entire team was furiously working to finalize the results we would soon make public. We had relentlessly reviewed the data, diligently debating the strength of the findings, discussing what could be one of the greatest scientific discoveries in history. In the intensely competitive world of modern cosmology, the stakes couldn't have been higher. If we were right, our detection would lift the veil on the birth of the universe. Careers would skyrocket, and we would be forever immortalized in the scientific canon. Detecting inflation equaled Nobel gold, plain and simple.

But what if we were wrong? It would be a disaster, not only for us as individual scientists but for science itself. Funding for our work would evaporate, tenure tracks would be derailed, professional reputations ruined. Once gleaming Nobel gold would be tarnished. Glory would be replaced by disappointment, embarrassment, perhaps even humiliation.

The juggernaut rolled on. The team's leaders, confident in the quality of our results, held a press conference at Harvard University on March 17, 2014, and announced that our experiment, BICEP2, had detected the first evidence of inflation—confirmation, albeit indirect, of the very birth pangs of the universe.

BICEP2 was a small telescope, the second in a series of telescopes located in Antarctica. I had co-invented the first telescope (BICEP) more than a decade earlier, when I was just a lowly postdoc at Caltech. BICEP sprang out of a deep obsession I had long had with making the previously inscrutable birth of the universe visible.

BICEP's design was simple. It was a small refracting telescope— a spyglass like Galileo's, with two lenses that bent incoming light and directed it not to the human eye but to modern, ultrasensitive detectors. The telescope needed to be at an exquisitely pristine location, and we found one: the South Pole. Our goal was to capture the aftershocks of cosmic inflation, a signal imprinted on the afterglow of the Big Bang—the CMB, which permeates all of space.

For years BICEP2 looked for a swirling, twisting pattern in the CMB that cosmologists believed could only have been caused by gravitational waves squeezing and stretching space-time as they rippled through the infant universe. What could have caused these waves? Inflation and inflation alone. BICEP2's detection of this pattern would be evidence of primordial gravitational waves generated during inflation, all but proving that inflation happened.

Then we saw it. There was no going back.

~

THE BROADCAST FROM Harvard's Center for Astrophysics captivated media around the world. Nearly 10 million people watched the press conference online that day. Every major news outlet, from the *New York Times* to the *Economist* to obscure gazettes deep within the Indian subcontinent, covered the announcement "above the fold." My kids' teachers had heard about it. My mother's mah-jongg partners were kvelling about it.

Watching the live video, I could see MIT physicist Max Tegmark reporting the event. He wrote, "I'm writing this from the Harvard press conference announcing what I consider to be one of the most important scientific discoveries of all time. Within the hour, it will be all over the web, and before long, it will lead to at least one Nobel Prize."

Finally, we'd seen what we, and the whole world apparently, had wanted to see. The BICEP2 team's announcement was that we had read the very prologue of the universe—which, after all, is the only story that doesn't begin *in medias res*.

Yet, in March 2014, I was just another spectator to the whole grand show. While I was still an official member of the BICEP2 team, I'd also partnered with members of another project, an experiment called POLARBEAR, designed to detect the same cosmic signals. Years before the dramatic press conference, my fealty to

BICEP2 had been questioned by its principal investigator, Harvard's John Kovac. An experiment's principal investigator is the scientific equivalent of a CEO. Kovac had determined that I was more competitor than collaborator; and now I was on the losing side. Like the dozen other experiments designed to detect B-modes from inflation, POLARBEAR was just another also-ran.

As BICEP2's four lead investigators took the stage at the press conference, I felt as though the discovery might go down in history with my name attached to it as a footnote, if at all. The pace and scale of modern science being what it is, scientists are lucky to have one shot at winning the Nobel Prize in their careers. I knew my best chance was BICEP2. I also knew I wouldn't be a part of it. A mixture of emotions flooded over me: joy and resentment, pride and jealousy, triumph and failure.

Still, doubts plagued me. It sure seemed to be a discovery for the ages. But was it? The famous physicist Richard Feynman said at the 1974 Caltech commencement, "The first principle is that you must not fool yourself—and you are the easiest person to fool." Cosmologists are keenly aware of this dictum, and we channel Woody Allen levels of neurosis to guard against confirmation bias—the tendency to see what one wants to see, to discard that which disagrees and favor that which comports with our own desired outcome. No one is immune from confirmation bias. And scientists, despite what you may think, are rarely mere gatherers of facts, dispassionately following data wherever it may lead. Scientists are human, often all too human. When desire and data are in collision, evidence sometimes loses out to emotion. We'd tried hard to heed Feynman's warning, but it was impossible to rule out every possible contaminant. Had we fretted enough?

The most worrisome aspect of BICEP2's signal was how huge it was. It was shockingly big, more like finding a crowbar in a haystack than a needle, as one team member phrased it. At the time of our announcement, we were worried about being beaten by our chief competitor, a billion-dollar space telescope called the Planck satellite with the perfect heavenly perch from which to scoop us. Prior to BICEP2's press conference, Planck had already ruled out a B-mode signal *half* as big as the one we claimed to have observed. Cosmologists were expecting a whisper. We claimed BICEP2 had heard a roar.

During the webcast, Harvard's servers struggled to keep up with the millions of hits pouring in from around the world. Through the stuttering live-stream from the Center for Astrophysics, I heard John Kovac say, "I'd like to highlight contributions from other major collaborators that you see in BICEP2 over the past years, including UC San Diego Brian Keating's group there. . . ."

Well, at least he mentioned me first. I had wondered whether they would mention me at all. Still, UCSD's logo was at the bottom of the PowerPoint slides. Only a week before, when they'd been prepared, my group's glyph had been at the very top of the PowerPoint slides, near the logos of the four other institutions leading the BICEP2 project. Now, obviously, in terms of effrontery, this was not exactly Putin's annexation of Crimea (which would happen the following day), but nevertheless, we were buried in the PowerPoint equivalent of the closing credits.

Did I deserve the demotion? I had bet on the wrong team, after all. For reassurance, I had called my friend Marc Kamionkowski, an astrophysicist at Johns Hopkins University. "There'll be other press conferences," Marc said to me on the phone the day before the press conference, as he made his way from Baltimore to Boston. "Not like this one," I moaned. On March 17, 2014, he sat alongside the four lead investigators, providing the media with an impartial interpretation of the highly technical findings. I was happy for him. His work had inspired me to create BICEP in 2001. But I wasn't in the inner circle anymore; I hadn't been for years.

The press conference continued for another hour, but the faltering of the live-stream servers proved prophetic. The ecstasy of the press conference was surreal. Not being there alongside past, and assuredly future, Nobel laureates was agonizing. I couldn't take it anymore. I went to my UCSD office to marinate in self-pity, alone.

The phone rang. It was Jim Simons, billionaire philanthropist, mathematician, and patron of the Simons Array project I was helping to lead. The Simons Array was supposed to find the signals BICEP2 had just found. Jim had been a colleague of my father and he'd become a mentor and friend to me over the years. He knew I'd created BICEP but was perplexed. Why wasn't I there? And why had BICEP2 beaten the Simons Array to the punch? "What's going on here, Brian?" he demanded in his gravelly Boston accent. What was going on, indeed?

Chapter 2

LOSING MY RELIGIONS

U.S. Amundsen–Scott South Pole Research Station, Antarctica

December 2005

"Dad's dying," said the crackling voice of my older brother, Kevin, over the phone, "but don't let on that you know." I couldn't breathe, let alone speak. I didn't have long; in a few minutes the satellite connecting us would vanish over the blindingly white horizon.

My first reaction was denial. He'd just gone in for a routine physical! After not living with our father for most of our childhood, Kevin and I had finally reconnected with him as adults a few years earlier. How could we be losing him . . . again? My disbelief was short-lived. It was quickly replaced by anger and bitterness.

Goddamn it. The timing couldn't have been worse. My entire academic career, my identity as a scientist, my dreams of winning a Nobel Prize, all were wrapped up in this massive cosmology project at the bottom of the world. The BICEP team had come to the South Pole to search for the beginning of time. As you might suspect, getting here had not been a quick jaunt. It had taken us four whole years to reach the Pole.

Antarctica is a treacherous place in the best of conditions. There is no way in or out for nine months of the year. In January 1912, the British explorer Robert Falcon Scott had arrived at the same place where I was now standing and famously declared, "Great God, this is an awful place!" Two months later he and four of his men froze to death on the sheet of ice that fanned out all around me.

I had come to the Pole with the tantalizing expectation that, by looking for evidence of creation, I might even find evidence of a Creator: God, either Scott's version or otherwise. At the very least, I hoped for a longer and more hospitable stay than the doomed Brits had experienced. After that phone call, it seemed like it would turn out to be neither.

I looked around the observatory. I'd been stationed here for a month, yet I still felt like a stranded astronaut on an otherworldly outpost. Eight thousand miles from home, suspended on two miles of snow, with buildings strewn about like houses on an infinite ice-covered Monopoly board, life had become surreal. Where all of Earth's time zones converged into one, I could find no marker for time's passage. The Sun's interminable brilliance made life outdoors blindingly unbearable. Even a summer day in the Sahara eventually yields to dusk. Not here, not now, not for months to come.

At the very point where the Earth spins on its axis, my world was spinning out of control. I saw the bright red panic button we'd put on BICEP's massive mount. In an emergency, you were supposed to press it to prevent the steel behemoth from crushing an unsuspecting astronomer beneath its whirling, multi-ton girth. A part of me wanted to push it, hoping that would stop time for me as well. But common sense intervened.

Ironically, though the outpost was dedicated to astronomy, the only astronomical body other than the Sun visible in the six-month-long polar "day" was the Moon. Glimpsing it through one of the observatory's thick portal windows was comforting. It was a familiar face, gracefully arcing above me, following me just as it had since I was a little boy. Unlike the cruel Sun, the Moon waxed and waned, changing its appearance, providing a crude way to mark time's passage. And so much time had elapsed since the Moon set me on a path of curiosity and discovery. The Moon had been guiding me all along. It was the reason I was here. It was the reason all of us were there.

DOBBS FERRY, NEW YORK
4 A.M., SEPTEMBER 1984

ARE YOU KIDDING ME, I left the light on . . . again? Clumsily stumbling to shut off my desk lamp, I stopped dead in my tracks. The light *was* off. The glow was coming from outside. I looked out the window. It wasn't a streetlight or a car; it was the Moon, and it wasn't alone. A mesmerizingly bright star was next to the plump full Moon, low on the horizon. What kind of star could rival the Moon? I was thirteen and I'd never seen anything like it.

For the next few nights, I watched the Moon and its brilliant companion arc across the humid Indian summer skies over Westchester County. I was perplexed. Google being fourteen years in the future, I had no choice but to wait until I got my hands on the Sunday *New York Times* with its "Cosmos" section.

After church services ended, Sunday's paper brought a revelation more exhilarating than any sermon I'd waved my altar boy's censer at. The "star" I thought I'd discovered was the planet Jupiter! Really, you could see an actual planet without a telescope or a spaceship? Another world, visible just with my naked eye. Imagine what I could see with a telescope!

I had to get one. But given my family's meager means, finding a spare $79 for the cheapest spyglass I could get would be a stretch. Fortunately, I had a job stocking shelves at the Venice Deli in Dobbs Ferry, making a cool $3.35 for each of the four hours of hard labor I put in each week. It was going to take a while: six weeks, to be precise. By then, summer would be long over and I'd be back at school, without any time to use the coveted new telescope.

Luckily for me, my scrimpings were supplemented by a generous grant from the first funding agency of many a budding scientist: my mother, Barbara. Soon, I was a principal investigator overseeing my very own 5 cm (2-inch) diameter refracting telescope. As soon as I got it, night couldn't come fast enough. Within hours, I saw Jupiter's moons, a retinue of four bright dots surrounding the king of the planets, reminding me of a jostling entourage surrounding a heavyweight boxer. Just like that, I had quintupled the number of moons I'd seen in my life. Later on, I saw the highest mountains and deepest craters on the Moon. After that, colorful stars of near infinite variety and many diaphanous "deep sky" objects such as nebulae and galaxies came into view.

That summer, I became a celestial evangelist, trying, mostly in vain, to persuade passersby to just look up. Showing the public visual delights with my first spyglass was a mixed bag. Some people got it. Most didn't. "Can we look at that neighbor girl again?" or worse, "You mean that white smudge in the eyepiece? Oh. That's it?"

When I needed something extra convincing, I'd call in the ringer: Saturn. Even the most jaded video game addict was left slack-jawed after glimpsing the magnificently disked, cream-colored world. But it wasn't visible every night of the year, and most people weren't overwhelmed by the other glowing blotches that entranced me. If you can imagine teaching music appreciation to a class filled with tone-deaf students, it was like that, only more disheartening.

Why didn't everyone delight in these ethereal skyscapes as I did? True, the telescope came with an ancillary benefit for me, as it transported me away from the troubles at home, the nightly arguments between my stepfather and my mother. Usually these were about money, or, more accurately, its absence. Our impending move, which would mean my fourth school in five years, was weighing on my mind. My telescope became my time machine, a ride to otherworldly vistas.

Soon I discovered that I needed more than a weekly dose of astronomy from the *New York Times*. I devoured every bit of printed astronomical media I could find at the library and kept meticulous logs of my nightly observations. The sights I saw were even more precious the more I read about them in the bible. Okay, not the actual Bible; it was the *Peterson Field Guide to the Stars and Planets*, which became my celestial scripture. I'd read the *Guide* all day, anxiously awaiting nightfall, a pimple-faced astro-Dracula craving the quiet shelter of the night. I still have my copy of the *Guide*; it's in my office at UC San Diego and proudly bears the autograph of its author, Jay Pasachoff, professor of astronomy at Williams College. Leafing through the *Guide*, I'm thirteen years old again.

The *Guide* contained extensive discussions of the planets and showed me how to observe them. For each planet, there was a description of how the great astronomer Galileo Galilei had perceived it when he became the first astronomer in history to use a telescope. Eventually, Galileo's observations led him to believe Copernicus's claim that the Sun, not the Earth, was the center of the solar system. Later, I learned that I'd unknowingly repeated Galileo's discoveries of four centuries earlier: the moons of Jupiter, the craters of the Moon, and the rings of Saturn. It was a connection that delighted me (as did the fact that he and I were both supported by Venetian patrons, mine being of the aforementioned delicatessen variety).

Galileo became my first hero. I read everything he wrote and everything that was written about him. I wanted to emulate him, to live out a modern version of his story in all ways . . . except that awful final chapter, when he was held under house arrest by the Vatican. That tale shook my faith to its core—an especially ironic twist given my own tortuous route to religion in the first place. Let me explain.

Although both my biological parents were born Jewish, they were completely nonobservant. We never went to temple or celebrated the High Holidays. Pork chops were on the menu not infrequently. After my parents divorced, my father moved to California. Growing up, I lived with my mother and my stepfather, Ray Keating.

In stark contrast to my irreligious upbringing, Ray had grown up in a deeply religious Irish Catholic home with nine brothers and sisters, and the whole brood treated us with the same love and affection as they would have if we had been born into their clan. When my mother married Ray, she, Kevin, and I converted to Roman Catholicism and were baptized. Soon afterward, my biological father allowed Ray to adopt Kevin and me.

I took to Catholicism immediately. The warm, solemn dignity and fellowship of Sunday services was captivating. Monsignor Robert Skelly's wisdom was rivaled only by his sense of humor. At age twelve, when most Jewish boys start preparing for their bar mitzvah ceremonies, I became an altar boy. Every Sunday, I assisted Father Skelly, taking great pride in dispensing Communion wafers with dexterity onto the tongues of the devout. I'd be lying if I said I didn't enjoy gently harassing the more frugal worshippers who attempted to ignore my masterful passing of the collection basket.

But my early experiments with Catholicism were inconclusive. Besides feeling morally unworthy of being a priest, as a teenage boy I had more than a little bit of angst over the prospect of remaining celibate for life. I knew I couldn't be a father in the Catholic sense.

It was right around then that I caught the astronomy bug and met Galileo. The more I learned, the more I thirsted to learn. Soon, I knew just enough to be to be dangerous. When I read that Galileo had been convicted of heresy by the Vatican, and forced—under penalty of death—to recant his evidence for heliocentrism as being "abjured, cursed and detested," I was deeply disheartened (after learning what "abjured" meant, at least). How could the Vatican threaten to torture him merely because of a scientific idea?

Even when I met Galileo, in 1984, the Church still hadn't formally pardoned him, despite three centuries of vindication. That put me over the top. Galileo was already my hero; now he became my idol, speaking to me across the generations. His creed was my rallying cry: "I do not feel obliged to believe that the same God who has endowed us with senses, reason, and intellect has intended us to forgo their use and by some other means to give us knowledge which we can attain by them."[1] Scientific means were the only means I needed now. Soon, I quit my post as altar boy. I didn't need a Father in Heaven; after all, I'd done just fine without a father on Earth.

For the next decade, that's where things stood. I was a devout atheist and proud of it. Scientific questions were all that mattered. I wanted to learn as much about nature as possible. So, at age seventeen, I became a physics major at Case Western Reserve University. After graduating, in 1993, I went to Brown University to pursue a PhD in experimental cosmology.

Although we cosmologists have healthy egos, being an experimental cosmologist doesn't mean you make and break universes. It means you study the cosmos by designing, building, and deploying advanced telescopes and detectors. As a graduate student, I got to work building telescopes that pushed the technology envelope— ultrasensitive detectors operating at mind-bendingly low temperatures. I was ecstatic. I got paid (albeit not much) to do what I loved: to build telescopes and to possibly answer the biggest question of all: how did the universe begin? But, I wondered, could a lowly graduate student really make an impact?

It wasn't long before I learned the answer to that question was yes. A month after I arrived at Brown, the winners of the 1993 Nobel Prize in Physics were announced. Half of the prize went to Russell Alan Hulse, who had been a twenty-three-year-old graduate student when he and his advisor, Joe Taylor, made the serendipitous discovery of a remarkable radio beacon, known as a pulsar, some 24,000 light years away from Earth. The pulsar Hulse and Taylor found was not alone; it had a companion, which caused its lighthouse-like pulses to slow down over time, ever so slightly. Hulse and Taylor showed that this remarkable "binary pulsar" was losing energy at a rate consistent with radiation in the form of gravitational waves, exactly as Einstein had predicted fifty-five years earlier. It was remarkable to think that a kid barely older than I was could have made a discovery for the ages. I wanted to do the same.

PAST IMPERFECT

We've all wondered: what if I could go back in time? What would I do differently? What if the things I did differently made my life *worse* instead? For now, traveling backward in time is impossible. Maybe that's a good thing; it's hard enough living life going forward.

These were questions my lab mates and I would ponder as we cooled our PhD experiments to unfathomably low temperatures. And while we knew that we couldn't ourselves go back in time, in theory time could be stopped, on a microscopic scale at least. To do so requires conditions unlike anything you or I have ever experienced, even at the South Pole: absolute zero temperature. What we call an object's temperature arises from the collective motion of all its atoms. In principle, one can cool atoms so that their motion ceases entirely. This occurs at absolute zero: zero kelvin, which corresponds to −273 degrees Celsius (−460 degrees Fahrenheit). For these atoms, motion—and maybe time, too—would cease; they'd float in a state of suspended animation.

Some say time, like temperature, is an emergent phenomenon, that it's only understandable in relation to motion. When I first learned about absolute zero as a beginning graduate student, I wondered: might time also have an origin? Just as the emergent nature of temperature was unknown until cryogenic technology arose to achieve its lowest values, so too would we learn more about time if we could make the right machine. Fortunately, we astronomers do have such devices: our telescopes. While light travels at extremely high speed, it can only travel a finite distance in a finite time. So, when you see an object at a great distance, you're not seeing it the way it looks "now," you're seeing it as it looked eight minutes ago, if it's the Sun, or 13.82 billion years ago, if it's the cosmic microwave background. But even the hoary old CMB light doesn't take us all the way back. For that we needed to build a very special telescope that could see gravity, in a way I'll soon explain. If we succeeded, we'd be looking back to the very beginning of time, or maybe even beyond.

In the first-year graduate student dorms at Brown, I shared a room with a foreign student named Tomás who'd come to America, in part, to study theater. But a more powerful motivation for him to move so far was to be close to his biological father. Like me, Tomás had been abandoned by his father as a child. When he was at college he'd reconciled with his father and, by grad school, the two had a close relationship. I admired how close they were, how they'd speak frequently, and how they'd even take vacations together. It was obvious that Tomás had given up his (in my view, legitimate) grudges against his old man. "That's why baggage comes with handles," he said, "so you can put it down." One day Tomás, knowing about my history, said, "You're studying the beginning of the universe . . . but you don't even know half of your own origin story."

His comment shook me. I'd lived so much of my life without my biological father being around that I'd convinced myself I didn't need a father at all. But by 1994, my curiosity about him could not be contained. I knew that my father, Jim Ax, was a mathematician and had been a professor at the State University of New York at Stony Brook. But I never knew what he actually did. I went to Brown University's science library and poked around. He'd probably studied something esoteric, something so stupefyingly dull that its only practical application was as a cure for insomnia.

Instead, I got the jolt of my life. His final papers before quitting academia were all about physics—my specialty—not math. And they weren't about "purely academic" subjects; they were on the biggest topics imaginable: the origin of time, the behavior of light, the nature of matter![2] Not only was I following in my father's academic footsteps, somehow I'd inherited his intellectual predilections as well.

I realized Tomás was right. I wanted to know where the seven-year-old's concept of my father left off and the adult's version began. I didn't even know whether he was still alive. But I decided to try to find him.

At the time, both of my biological grandmothers lived in the same zip code in south Florida. To badly mix metaphors, Sunrise, Florida, was a veritable mecca for Jewish grandmothers. So, I asked my mother to ask *her* mother, Lilian, whether my father's mother, Esther, might be interested in talking to me. When my father got wind of these inquiries, he called me in my dorm at Brown. Despite not having heard his voice in over fifteen years, I recognized it instantly. "This is Jim Ax," he declared in a thick Bronx accent.

We spoke for hours. He had retired from academia and had remarried. He was living in Los Angeles and seemed happy, though it was obvious he was still stinging from the divorce and adoption, even though he'd willingly agreed to both decades earlier. Try as he might, he never was able to suppress the parental love, or at least the curiosity, he had for his sons unseen. He blamed his decision to give us up for adoption on my mother. But I didn't buy it; she was, and is, the most incredible person I've ever known. Still, we agreed to keep the lines of communication open. And speaking of open, every time we'd speak, my sides would nearly split.

Despite quitting academia, he hadn't given up on science. He'd converted from math to physics, investigating the nature of matter and energy at subatomic scales with his longtime colleague, Princeton University math professor Simon Kochen. He was thrilled that his offspring was studying cosmology, trying to answer the biggest questions imaginable, even if it was only his genes, not his fathering, that had compelled me to do so. Soon after that first phone call, Kevin and I reunited with our father. Eerily, we picked up as if nothing had happened.

Over the next few years, the three of us made up for lost time. In my case, we developed a relationship in many ways much closer than that of a typical twenty-something and his father. This he attributed to the freshness of our relationship: "I wasn't around when you were growing up. Now, I get to experience the good stuff, you know, the non-diaper-changing stuff." We shared a lot of ideas through my six years of graduate school, years I spent learning how to build telescopes to mine the aftershocks of creation. After I defended my PhD thesis in 1999, my father said, "I can't take credit for your success, because I didn't raise you. Your mother gets all the credit for that." But, he said, "I'll gladly take credit for your genes, if only half of them." With our reunion came a second chance for him to participate in my life. It was an opportunity neither of us wanted to blow.

After leaving Brown in 1999, I moved to Stanford University. In part, I chose Stanford because I wanted to work with a brilliant young physics professor there named Sarah Church. She was working on a new microwave telescope, to gaze at distant galaxies. But another reason for the move west was to be closer to my father.

THE WINDS OF FREEDOM BLOW

Life in Palo Alto in September 1999 was miserable for a postdoc making $35,000 per year with college loans to pay off. My mood moved inversely with the late 1990s' NASDAQ. The booming stock market severely limited my housing options.

The only apartment I could afford was miles off campus, located on the major Caltrain artery connecting the Valley to the City, at the precise location where the train conductors were required to blow a 150-decibel horn to warn would-be rail-jumpers of impending doom. The horns began blaring at 5 a.m. It is said that, as a postdoc, Einstein was inspired by the passage of trains along a platform to develop new concepts of light, energy, and matter. Luckily, Einstein didn't live near the Caltrain line; had he, we'd still be waiting for $E=mc^2$.

On a good night, I would accumulate about five hours of sleep. I was exhausted and depressed. My postdoc advisor, Sarah, could sense it. When I'd doze off at work, I'd dream about a new type of telescope, which would later become BICEP. In my mind's eye, it was a telescope that could see all the way back to the Big Bang. I didn't share my ideas with Sarah. She had her own telescopes and a little thing called tenure to worry about too. After four years as an undergraduate, followed by six years as a graduate student, I craved autonomy. I wanted the freedom to work on my own ideas, to solve my own problems, to pursue my own astronomical aspirations.

For the first few months of my time at Stanford, I was completely absorbed by a new paper called "Polarization Pursuers' Guide," written by cosmologists Andrew Jaffe, Marc Kamionkowski, and Limin Wang.[3] Like Pasachoff's *Field Guide* sixteen years earlier, this "Guide" became another guide for the perplexed. It enraptured me. It was the first time I heard anyone say it was possible to *experimentally* probe the first instants of cosmic history, that mysterious epoch called inflation, which I promise to explain fully in chapter 7.

Not only was it possible to see whether inflation had really happened, but, according to these cosmologists, it would require only a modest-sized telescope. I had worked on just such a small telescope for my PhD project.[4] I knew that tiny telescopes come with big bonuses: they are simpler, more efficient, and less expensive than their bigger brothers. A small telescope that could capture microwaves, rather than optical light like Galileo's, could probe the inflationary epoch just as well as a telescope many times its size and cost. It seemed too good to be true.

Then, one day, a mere six months after I defended my PhD, as I was lost in thought again, Sarah Church walked into the lab and told me she was unhappy with my performance, my attitude, and my work ethic. For the first time in my life, I was fired—my career ended before it even started. I couldn't help but think this was yet another ironic parallel with my hero, Galileo: both of us had been on the wrong side of the Church. I couldn't argue with Sarah. I had it coming. The months I'd spent fantasizing about new telescopes were months I should have been working on *her* projects.

I was ashamed and dejected. I had moved west to make my father proud, and now it seemed likely that I'd be living under his roof for the first time in twenty years, sleeping on his couch while my stepmother harangued me to get a job. At that moment, the prospect seemed scarier than facing the Inquisition.

After Sarah left the lab, I looked at my lab notebook. It had

Stanford's tree logo on its cover, and underneath, Stanford's motto: *Die Luft der Freiheit weht,* which means, "The winds of freedom blow." Well, this blows, I thought. But wait—I was free, too; with no job, I was unencumbered by obligations (if also liberated from a paycheck).

Before my termination, I had been working on someone else's experiment. Now I could focus on my own ideas. I'd need a lot of support, though. Convincing yourself that you're brilliant is easy. It's much harder to sway others to see it so. But science was on my side; I knew it was possible, in principle, to experimentally measure the cosmos's initial conditions. It would require more than just raw sensitivity. It would need an amazing team, an ingenious design, and an exquisite location.

Sarah provided a generous severance package. She kept paying me for a while, and she put me in touch with her former postdoc advisor, Andrew Lange, an experimental cosmologist at Caltech. A few weeks after firing me, in June 2000, Sarah told me that Lange was coming to speak at Stanford. I was sure it was to be a preview of his future Nobel Prize acceptance speech.

Why the great expectations? In April 2000, the BOOMERanG experiment Lange led did something phenomenal.[5] Lofted by a helium-filled balloon into the upper stratosphere, BOOMERanG had detected a telltale pattern in the ancient CMB light which revealed the age, composition, and structure of the universe more accurately than ever before. BOOMERanG was as close to a slam-dunk Nobel Prize–worthy experiment as I had ever seen. It worked perfectly the first time. Led by a brilliant and charismatic principal investigator, BOOMERanG had solved what was, literally, the biggest geometry problem in history. It measured what is known as the spatial curvature of the entire universe.

There was electricity in the air when Lange spoke at Stanford in June 2000. I had never heard a scientist like him: part Steve Jobs, part Tony Robbins. He was calm, forceful, and mesmerizing as he speculated on future directions for the field of cosmology. He began

FIGURE 1. A spiderweb bolometer detector, like the ones BOOMERanG used to measure the universe's curvature. Photons, particles of light, from the cosmic microwave background are absorbed by the spiderweb structure, causing the web to warm up. The rectangular thermometer at the center of the web measures the temperature rise, telling cosmologists how much microwave energy landed on the detector. (NASA)

with an elegant description of how his delicate bolometers (from the Greek for "radiance meter") worked. Bolometers are detectors that measure electromagnetic radiation by turning it into heat and then measuring the temperature rise by converting the heat to electrical signals. BOOMERanG's spiderweb bolometers (fig. 1) used a fine mesh to absorb the CMB light, which, as the name implies, is in the form of microwaves. Microwaves, like light waves, are a type of electromagnetic radiation; their wavelengths are a few thousand times longer than the wavelengths of visible light.

These bolometers, as Lange described them, were designed the way Mother Nature—the greatest architect of all—would make them. A spider minimizes the amount of biological energy it expends on spinning its web, using just enough energy to catch the flies it needs to survive. It doesn't spin a solid disk—that would surely catch everything, but at enormous biological cost to the spider. Nor does it frugally weave just a few strands; too many flies would slip through. The spiderweb bolometers were built, Lange said glee-fully, with exactly the right size weaving for the prey BOOMERanG was hunting—microwaves from the Big Bang, each with an aver-age wavelength of two millimeters. He was a genius storyteller. The crowd was enraptured.

From BOOMERanG, cosmologists learned that the universe is flat, meaning that if you draw a triangle between three stars, three galaxies, or three of any extraterrestrial object—no matter how big the triangle is, the interior angles always add up to 180 degrees, just as those angles did on papyrus for Euclid millennia ago. This dis-covery was astounding, as cosmically consequential in our time as Eratosthenes's proof that the Earth was not flat was in 200 BCE. BOOMERanG's findings were vital evidence for the theory of infla-tion, which predicted that the universe should be flat, because the primordial hyperexpansion of space would completely flatten out any curvature, however significant it had been at the beginning of time. (Exactly how inflation accomplishes this immense ironing of space-time will be described in chapter 7.)

When Lange and his team released their astonishing results, cos-mologist Michael Turner of the University of Chicago called it "the day cosmology changed."[6] Although evidence pointing toward a sim-ilar conclusion had been presented (with far less fanfare) six months earlier by a team led by Princeton physicist Lyman Page, many scientists considered BOOMERanG's findings more convincing, claiming they confirmed the theory of inflation.[7] But, Lange said, BOOMERanG didn't actually *prove* that inflation had occurred. BOOMERanG's remarkable results were circumstantial evidence at best—certainly compelling, but not proof. In fact, he went on, prov-ing inflation beyond a reasonable doubt could never be done with the type of measurement BOOMERanG had made.

After his talk, Lange agreed to chat with me for a few minutes. I had heard so much about him I felt like I knew him. He was forty-two years old and had been at Caltech since 1993, after a mete-oric rise from freshly minted PhD in 1987 to professor, both at

UC Berkeley. Caltech had been courting him for a while, betting that his stock would continue to rise. BOOMERanG proved they were right. By the time I met him, he was rumored to be the most popular professor at Caltech; the lecture I'd just heard confirmed those rumors.

Six feet tall, wearing an Oxford shirt with its top two—not one—buttons undone, Lange reminded me of a 1950s adman. He had a piercing gaze that made me feel he was completely focused on me when I spoke, which was disarming. He shocked me by saying he had read my doctoral thesis and thought it was "masterful." Of course, I accepted his compliment without protestation. He asked me about my plans after Stanford, avoiding the elephant in the room: my untimely termination. "Why don't you come to Pasadena and give a talk in a few weeks?" he asked.

I arrived in Pasadena one month later and nervously practiced my talk in a seedy motel a mile from Caltech's manicured campus. I knew this was my one shot at impressing cosmology's greatest show-man. I gave it my all, pitching the dream telescope I'd been obsessing over for the past year. Speaking to Lange and his group, I described a novel CMB telescope based on the teachings of the "Polarization Pursuers' Guide," a glass that could spy all the way back, even beyond BOOMERanG's startling vistas. I casually mentioned that the "Guide" implied that a small telescope could see just as far back as a much larger, and therefore more expensive, telescope. It was like BOOMERanG's elegantly minimalist spiderweb bolometers: build only what you need to get the job done. Lange seemed impressed. After the talk, he offered me a job. I accepted before he finished the sentence.

Two months later, I got to work on the experiment that would become BICEP. I resolved to earn the luck I had clearly chanced upon. Like the other six postdocs in Lange's lab, I worked hard to impress him. With his enthusiasm, renown, and brilliance, and with our commitment to his cosmological enterprise, his group was a force to be reckoned with.

Lange hosted social events at his house and clearly enjoyed his role as both a mentor and a friend. Soon, he and I developed a close bond. Often, he'd give me nuggets of what he called "fatherly advice": words of wisdom about science, academia, and occasionally even about actual, biological fatherhood. The latter information, while not immediately relevant, made a deep impression on me nonetheless. He worshipped his three young sons. His office was a museum of their artwork. On the shelves he'd placed their science projects right next to his own award plaques and pieces of the actual rockets he'd launched into space.

Most Mondays, he'd regale me with tales of his weekend exploits with his sons, camping out or launching model rockets in the Mojave Desert. It was clear that they were his world, a refreshing revelation to me: you could be one of the world's most brilliant scientists and still make fatherhood your top priority. I appreciated his wisdom more than he knew, for I had kept hidden from him my long estrangement from my own father.

Four years later, under Andrew's stewardship, I found myself at the bottom of the planet and on top of the world of science. We had funded and built BICEP, the experiment I had long dreamed about, the first experiment that could take us back to the beginning of time. But then, almost as soon as I had arrived at the South Pole, in late 2005, time stopped altogether.

DECEMBER 2005

"HOW MUCH TIME does he have?" I asked my brother, Kevin. "I don't know," he said. "You'd better get back here soon." The phone connection cut out as the satellite vanished over the icy horizon. I was scared and alone.

Staring at the hulking blue and white BICEP telescope, I flashed back to my childhood telescope. I realized that it, like BICEP, was a refracting telescope, a telescope that used lenses to bring distant sights into focus, and in that sense it was just like the spyglass through which Galileo had peered at the skies above Padua centuries earlier. Only a few days before BICEP had gotten "first light"—a milestone for any telescope, marking the first time it officially collects light from the heavens. The adventure was just beginning. But for me, it was ending.

I thought back to how much had changed for me, from my first telescope to BICEP. Back then, I had no inkling of how my life was going to turn out. I just knew it would be guided by science, not faith. Back then, I wasn't even talking to my father. Now, after rekindling the relationship, I didn't know how many conversations we might have left.

> *"Honor thy father and thy mother: that thy*
> *days may be long upon the land."*
>
> EXODUS 20:12

DESPITE HIS ATHEISM, Dad was quite fond of this dictum, the Fifth Commandment. He would often issue it to me as part of an order to get him something from the fridge. Holding up the five fingers on his open hand, he'd jokingly threaten me with bodily harm if I didn't obey him. "Remember the Fifth!" he'd shout shrilly, and I'd respond with feigned fright.

I had begun exploring my Judaic roots only a few years before arriving at the South Pole. To break the negative patriarchal cycle that my father, and his father, had started, I looked for role models, other father figures, everywhere I could find them. I even turned to long-dead ones like the biblical patriarchs. My father reluctantly resigned himself to my newfound exploration into the faith of my birth, saying, "I don't believe in God, but I *do* believe in the Devil because he made you believe in God."

In my teenage days, when the Church's treatment of Galileo shook my faith, I might have agreed with him. Now, with decades of life experience and some wisdom as well, I wanted to be a bit more methodical in my exploration into faith. As a scientist, I believed that every factual claim can be assessed using the tools of experimental science. Why should religion be different from studying the cell or the cosmos? Scripture should be accessible to proof or, at the very least, falsification.

Naturally, I set out to devise an experiment, hoping it would not turn out to be a fool's errand. I thought of ways to experimentally examine the Bible, probe its professions, check its claims, and subject it to science. I wasn't the first to do this, of course; I had long known about the theological gamble known as Pascal's wager.

In the seventeenth century, the French philosopher Blaise Pascal famously devised a way to test two contrasting beliefs: God exists/God doesn't exist. Under each possibility, Pascal described the two behaviors one could possibly take, and teased out a cost–benefit analysis of the four possible options. For example, the penalty for *not* believing in God if God *does* exist was eternal damnation—arguably even worse than graduate school. But Pascal's wager was based on the minimization of suffering; he wasn't concerned with the underlying truth or falsehood of the opposing hypotheses. And Pascal's reward for obeying God, if God exists, was only in the hereafter. How could I test it in the here and now?

So I set out to identify biblical predictions and subject them to falsification. If the claims were disproven, I would reject Scripture too. But what divine data set could I use to test the Bible? Then I realized that, unique among the Ten Commandments, the Fifth Commandment seemed to offer reward here on Earth. How could everyone have missed this?

How interesting it is that a tome revered (and reviled) for its imprecations and moral prescriptions offers a way to test its assertions. The reward of the Fifth Commandment—the lengthening of one's days upon the land—was especially appealing to me as a physicist. That concept offered a real-life form of Einstein's time dilation. But, whereas Einstein required acceleration to near light-speed to slow the passage of time, the Fifth Commandment gave me an opportunity to put my fledgling faith to the test: by honoring my father, taking care of him, comforting him in irretrievable days that could prove to be his last, I believed I could experimentally examine the Bible. I was sure I could call the Bible's bluff.

From the Pole, I called Andrew Lange and told him about my dad's diagnosis. He knew how much BICEP meant to me, but he knew family mattered more. "Brian, you've got to drop everything for your father," he implored.

I knew he was right. But still, it was frustrating on many levels. My trip to the South Pole was supposed to be the biggest quest of them all, a race to the beginning of time and a journey so manifestly significant that, if we succeeded, a Nobel Prize would surely follow. My father was sure to be impressed and, yes, to be proud of that as well.

The irony of it all was painful. For years I'd dedicated my life to trying to reveal the cosmos's origin story, even though my own version was mostly a mystery. Then, suddenly, just when the telescope I'd "sired" had reached adolescence, I was forced to abandon it, just to comfort a man who'd left me as a child. In my gut, I knew there'd be many telescopes to come, but I only had one father, flaws and all.

So, in December 2005, at the height of my Antarctic adventure, my days on the South Pole's icy landscape were cut short. Even though I had just arrived, I was about to head back north. Of course, from where I was, all directions faced north. I just needed to follow the invisible meridian connecting him to me, somewhere beyond the far white horizon.

Chapter 3

A BRIEF HISTORY OF TIME MACHINES

THE JOURNEY WE'RE ABOUT TO EMBARK ON WILL take us back to the beginning of our universe. But reading this story requires a very special pair of glasses, a set of lenses first used by none other than Galileo Galilei on a balmy June night in 1609.

That evening, Galileo did something no one had ever done before: he looked up, through a tube that was destined to become a biblical bludgeon. He had come to the tube in a rather nefarious way, committing academia's cardinal sin: plagiarism. His friend Paolo Sarpi had recently told him about a new device invented by a Dutch spectacle maker that could make distant objects appear closer. Remarkably, even though he had not seen the Dutchman's device, Galileo not only reproduced it but built several more with vastly improved performance. Soon, his *perspiculum* ("perspective tube" or "spyglass") would become the shortest lever ever to move the Earth.

It was a simple invention, really: two glass lenses within a flimsy lead tube. Each lens had one flat side. One of the lenses was bowed out on one face, like a squashed lentil; indeed, the diminutive legume's Latin name is the origin of the English word "lens." The other lens's reverse face was concave, bowed in. Each lens was about 5 cm (2 inches) in diameter. The one closer to the object under study, the objective, was at arm's length from the eyepiece, which, unsurprisingly, is where you put your eye. The combination of those lenses, placed the perfect length apart, magnifies distant objects.

In the span of one remarkable week in June, Galileo perfected the device. Soon afterward, he increased its magnification to thirty-power, reduced troublesome optical artifacts, and added an adjustable tripod to stabilize the scope, since its magnification also amplified the wobble of the user's hands. The Dutchman's device lacked these features, rendering it a mere curiosity. In contrast, Galileo's telescope (from the Greek for "distance viewer") was a serious scientific instrument.

Driven by mankind's second oldest impulse, Galileo immediately realized the telescope could also make him rich. This was no small motivation for Galileo, for he had several illegitimate children and jilted ex-lovers to support. For a while, he had a monopoly. He could sell the new and improved spyglass to the highest bidder—and the bidders were manifold. His telescope provided a tremendous military advantage. With it, a naval commander could spot an enemy flotilla as soon as it was over the horizon, rendering null the stealth that distance had previously afforded.

His first concern was to maintain his monopoly. Galileo trusted no one. He closely guarded his design until he was ready to go public and claim credit.[1] In a Renaissance version of the television show *Shark Tank*, he demonstrated his telescope to prospective patrons. Eventually, he pitched it to the Venetian doge and senate. They bought it—several copies, in fact—and granted him a lifelong lectureship in Padua. They also doubled his salary. (As a professor, I'm awestruck by Galileo. Not only did he instantly secure tenure, but his department chair also doubled his startup funding.)

But it is not the world's shortest tenure track that we professional astronomers envy most. Galileo was the first person to unleash the full power of telescopic vision, making him privy to all sorts of low-hanging fruit on the Tree of Knowledge. Imagine having your own personal Large Hadron Collider—the world's most powerful particle accelerator—before anything like it existed elsewhere. (I admit, this machine is not at the top of most normal people's wish lists.) Nevertheless, imagine you had the time and power to use it as you wanted. How easy it would be to discover every subatomic particle, scooping your competition while you're at it!

Of course, the sharpest tools become blunt and impotent in the hands of the ignorant. Desire and passion are needed to take advantage of your talent and tools too. It's why cosmologists spend four years as undergraduate physics majors, and then at least a decade as graduate students and postdocs. We thirst to acquire both the knowledge and the wisdom to maximize the potential of these powerful tools—be they telescopes or supercomputers, or even our own brains.

MISTAKES, I'VE MADE A FEW

Fortunately for history, Galileo had more than just the best technology; he was abundantly curious, tireless, methodical. He was both humanity's first physicist and its first observational astronomer. Some historians go further, calling Galileo the first true scientist— the first to employ the scientific method, the iterative technique of fact-gathering and model refinement.

Unlike Newton, who dabbled in alchemy and other dubious endeavors, or Aristotle, whose "laws of nature" were, essentially, all incorrect, Galileo's reputation is also one of infallibility, an aura which is, in part, due to self-selection: Galileo didn't publicize his mistakes. He was also a skillful self-promoter, never failing to assert the priority of his discoveries, in case they might be monetized.[2] And he was supremely competitive. So concerned was he with maintaining his monopoly, Galileo even hid his discoveries from his contemporary, the German astronomer Johannes Kepler, who was the first scientist to accurately describe how the planets move.

In his landmark work, *Sidereus Nuncius* ("Starry Messenger"), Galileo wrote, with boundless confidence, "Surely it is a great thing to increase the numerous host of fixed stars previously visible to the unaided vision, adding countless more which have never before been seen."[3] The maestro never feared to push the boundaries of his theories. In hindsight, he should have been more cautious.

Still, his initial grandiose claims are forgivable; not only was publicity the main means of asserting priority, he was also rightfully proud of his accomplishments—achieved despite astronomy's inherent inadequacy compared to other scientific disciplines. For astronomy, it must be said, cannot really compete with other experimental sciences. Even modern-day astronomers cannot do true cause-and-effect experiments: tests using variables that change according to our will or controls that we can keep in quarantine, perfectly isolating the effects of the variable factors. Astronomers cannot create a black hole in the lab, alter a few conditions, watch what happens, and then compare to reality. Rather, we astronomers must make do with what arrives in our telescopes, carefully scrutinizing the cosmic flotsam and jetsam that rides upon electromagnetic waves to our shores.

THE GREAT DEBATES

Things were even worse before Galileo. Back then, astronomy was a mostly qualitative exercise. Having been birthed from astrology, the practice of divining future terrestrial events based on celestial phenomena (then, as now, not exactly known for painstaking peer review), astronomers had to use the limited data they had. However, these technologically challenged astrologers performed a valuable service: they accumulated archival data that, in the hands of Kepler and Galileo, would upset the orthodoxy of their age and initiate the first of what I call the Great Debates.

The Great Debates are arguments throughout the centuries that revolve around the question of whether human beings are truly a privileged class of observers, located at a particularly propitious time and/or a fortunate place within the cosmos. In the early

1500s, the Polish astronomer Nicolaus Copernicus initiated the first debate when he postulated that the solar system (well, what we now call the solar system), which was at the time considered to be the entire universe, was focused on the Sun.[4] Copernicus demoted the Earth from center of the universe—the most auspicious of all locations—to a humbler station as a mere planet, joining the five others making laps around the Sun. His theory was astounding and a slap in the face of Ptolemy, the second-century astronomer whose geocentric model, the idea of an Earth-centered universe, had prevailed for more than a millennium. Posthumously, Ptolemy obtained a collaborator—the Vatican—which held that Scripture implied all things indeed revolved around the Earth.

In January 1610, after sketching the Moon, Galileo set his sights on the two brightest itinerant drifters of the night sky, Venus and Jupiter. (Our word "planet" stems from the Greek word for wanderer.) What he saw astonished him. Under his telescope's magnification, each planet's appearance *changed* over time. He saw that Venus mimicked the Moon's cycle—the crescent he saw at first morphed, a few months later, into a plump, gibbous shape. He reasoned that the only way this alternation in shape could take place was if Venus orbited the glowing Sun, not the Earth. This behavior was impossible according to Ptolemy. It was a revolutionary idea, although it didn't belong to Galileo; Copernicus had theorized this decades before him. But it was only because of Galileo's data that Copernicus's ideas could be elevated from conjecture to fact.

If a picture is worth a thousand words, Galileo's next target would fill a metaphorical dictionary. The most convincing evidence in favor of Copernicus came when Galileo turned his telescope to Jupiter. In doing so, he made the first stop-motion animation in cosmic history (fig. 2).

Through Galileo's telescope, Jupiter was revealed to be its own microcosm, a system within *the* system. Jupiter had its own wandering vassals: four tiny points of light, which Galileo called stars,

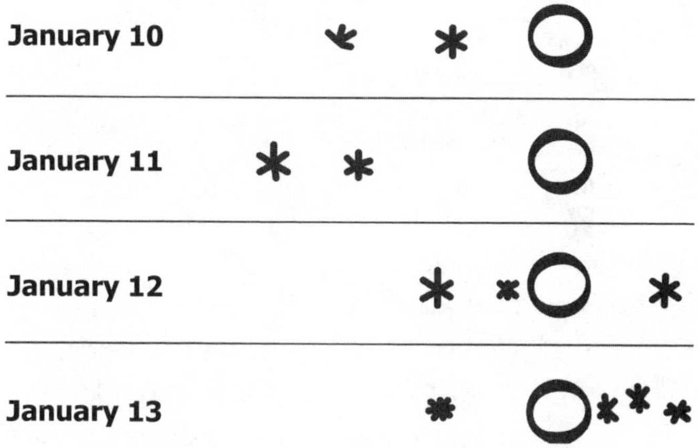

January 10			

January 11			

January 12			

January 13			

FIGURE 2. A reproduction of Galileo's 1610 sketches of Jupiter and its four moons that were visible in his telescope. He deduced that the moons' orbits were periodic by watching them over many nights. These observations would later become the most powerful evidence against Ptolemy's geocentric model of the universe. (© SHAFFER GRUBB)

sycophantically naming them after his patrons, the Medici. These were Jupiter's moons and their clockwork orbits were all clearly centered on Jupiter, not Earth.[5] He tracked the movement of what he called the Medicean stars for several days in January 1610 and made a somewhat repetitive silent movie that nevertheless had the star power to move the entire Earth.

His Jovian observations were an obvious refutation of Ptolemy's geocentric universe. Not only was the universe dynamic, but the Earth was no longer the sole focus of motion (fig. 3). Though Galileo did not *prove* that the Sun was the center of the universe,[6] he had clearly falsified geocentrism, ending the first Great Debate, upsetting Ptolemy (and the Church) and corroborating Copernicus. Galileo's refracting telescope became the first lever to displace our planet from the centrality it had enjoyed since antiquity. This was the first piece of evidence for what later became known as the Coperni-

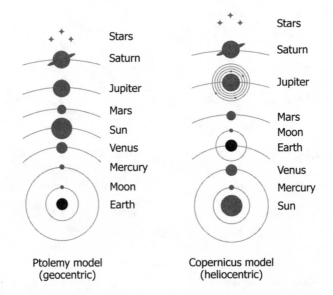

FIGURE 3. Ptolemy's geocentric model of the universe compared with the Copernican heliocentric model. (© SHAFFER GRUBB)

can principle, which asserts that there are no privileged places in the universe.

> *"Do not follow the desires of your heart
> and eyes that lead you astray."*

NUMBERS 15:39

RIDING HIGH, Galileo turned his attention beyond the solar system, once again seeking evidence to bolster the Copernican theory. In the winter of 1610, he trained his telescope on the Pleiades, a clumpy cluster of stars in the constellation Taurus. He perceived far more than a mere seven stars—the source of their common name, the Seven Sisters. Instead, an uncountable array of stars came into view, most of which were invisible without the telescope. Mysteriously, he saw the stars were ensconced in a milky bluish radiance. What was causing it?

FIGURE 4.
The Pleiades as sketched by Galileo in *Sidereus Nuncius* (*top*). Galileo described the cloudy glow between the naked-eye stars as the "nature of the Milky Way" itself, mistaking the dusty reflection for yet more unresolved stars. *Middle*: the Pleiades seen with the Hubble Space Telescope. *Bottom*: close-up of the star Merope, revealing the dusty reflection nebula. (*TOP*: FROM THE COLLECTION OF JAY PASACHOFF; *MIDDLE AND BOTTOM*: NASA/ESA/AURA)

Galileo, the polymath paragon of the Renaissance man, was not only a scientist. He was also an artist. All great artists have muses, and Galileo had seven: the Pleiades. The blue glow suffusing the Sisters transfixed him. It was as if luminescent blue paint had been spilled on the inky black velvet of the night sky. He couldn't explain it. But that didn't prevent him from hypothesizing about it. In early 1610, he made a sketch of the Pleiades that illustrates both a superhuman intellect and an all-too-human bias (fig. 4).

The *Sidereus Nuncius* combined Galileo's telescopic observations with brilliant new hypotheses. His depictions of the Jovian satellites in the *Sidereus Nuncius* were the most convincing evidence against the Earth-centered universe. Galileo should have stopped there. Instead, his telescope gave him tunnel vision, depth without a full perspective. For the remainder of his career he sought out ways of confirming Copernicus, rather than simply disproving Ptolemy.

His Copernican campaign continued in the *Sidereus Nuncius* with a claim that the fuzziness that surrounded the Seven Sisters was caused by the combined glow of innumerable stars, an assertion that what had previously been thought to be empty space between the stars in the Pleiades and the rest of the Milky Way was filled with "invisible stars" (invisible, that is, to the naked eye). And, not surprisingly for the first astronomer to wield a telescope, when it came to the five senses, to Galileo sight was first among equals. Later he would say his observations deprived "former writers of any authority, since if they had seen what we see, they would have judged as we judge."

By induction, Galileo reasoned that he could see more and more stars if only he could increase his telescope's magnification. "What was observed by us [. . .] is the nature or matter of the Milky Way itself, which, with the aid of the spyglass, may be observed so well that all the disputes that for so many generations have vexed philosophers are destroyed by visible certainty, and we are liberated from wordy arguments," he wrote. (Then, as now, some physicists just couldn't resist upbraiding philosophers.)

Galileo's revolutionary proposition was that the Pleiades (and other so-called nebulae) were comprised entirely of an uncountable assemblage of stars which blended together to produce a cloudy glow. Not only that, but such was the composition of the *entire* Milky Way galaxy. The Milky Way's milkiness came from the blending of innumerable unresolved stars. What Galileo had done to the Earth, he now did to the Sun: it was demoted to just another star among a myriad of myriads in the Milky Way, a bit player in an essentially infinite ensemble cast. Humanity's modesty wasn't confined to just our solar system. It reached to the stars.

We know now that the milkiness of the Pleiades (and other nebulae like the Orion nebula) is almost entirely made of dust and gas. Nebula dust is remarkably rich—more like soot than the dust you find on your furniture. In addition to iron and nickel compounds, it contains copious quantities of pure carbon—in some cases, diamond dust, just as we all learned in the nursery rhyme "Twinkle, Twinkle Little Star." The Pleiades later became the paradigm of the nebular class known as reflection nebulae, stellar nurseries harboring great quantities of astronomical afterbirth, their gas and dust reflecting the eerie bluish glow of the infant stars swaddled within. No matter how much magnification is used, reflection nebulae like the Pleiades and the Orion nebula will never be resolved into stars— because the glow isn't coming from stars.[7]

Galileo was the victim of a dusty delusion. But dust didn't do it alone: it had an accomplice—desire. Confirmation bias preyed on Galileo's impulse to further buttress the Copernican model of the universe.

CAUGHT BETWEEN THE MOON AND VATICAN CITY

Sidereus Nuncius made Galileo an overnight scientific celebrity. But the deflation of humanity's cosmic pride would come right before his fall. His tenured post at Padua brought him closer to the Vatican and the Holy See, the executive branch of the Inquisition. And though the Medici moons won him favor with his patrons, his tact deserted him over the years as he continued to explore the heavens with his telescope.

After decades of observation and contemplation, Galileo fully fleshed out what he considered incontrovertible evidence for heliocentrism. In his book *Il Dialogo sopra i due massimi sistemi del mondo* ("The Dialogue Concerning the Two Chief World Systems"), he chose to settle the science of the Sun-centered cosmos by imagining a conversation between a proponent of the biblical view of fixed-Earth cosmology named Simplicio (as in "simpleton"), who bumbles his way through complex Aristotelian arguments, and Galileo's alter ego Salviati, whose intellect outshines the very astronomical objects he describes. The pitting of science against religion in this way was bound to catch the attention of the once friendly Pope Urban VIII. The crucial pro-Copernican argument in the *Dialogo* concerns Galileo's theory of the origin of Earth's ocean tides, which, he claimed, resulted from the Earth's motion, a sort of sloshing that occurred as the Earth rotated on its axis and orbited around the Sun.

In truth, this was a bigger blunder than his Pleiades conjecture; the tides are due to the Moon's gravitational influence, not to the revolution and rotation of the Earth. Like his hypothesis about the purely stellar composition of the Milky Way, Galileo's incorrect theory of the tides is another example of confirmation bias corrupting his otherwise brilliant arguments. His starting point was Copernicus, and everything had to flow from it, even the tides.

Either way, espousing heliocentrism had been banned by the Church since 1616. Pope Urban could no longer tolerate the scholar's blasphemy. In 1633, Galileo appeared before the Holy See's kangaroo court and was summarily imprisoned. Despite his alleged, indecorous mutter of "And yet it moves," the Pope graciously allowed Galileo to live out the final decade of his life near his illegitimate daughter in a hilltop villa overlooking Florence. I suspect that the maestro would be most depressed, however, by a study—ironically released by the U.S. National Science Foundation on the 450th anniversary of Galileo's birth—which indicated that in 2014, a quarter of Americans, and a third of Europeans, were not aware that the Earth revolves around the Sun.[8]

CONSOLING OUR COSMIC EGO

Galileo and the Church set the stage for many more Great Debates. Just because the Earth wasn't the center of the solar system, that didn't mean that the solar system itself wasn't in a privileged place.

Galileo died in January 1642. Fortunately, the continuum of great physicists endured, for Isaac Newton was born on Christmas Day that same year. Amid his explorations of alchemy and the occult, Newton found time to invent a new type of telescope, one that used curved mirrors instead of lenses to collect and focus light. The advantage of the reflecting telescope was that large-diameter mirrors were easier to make than large lenses. Therefore, reflecting telescopes have a much wider collecting area, which in turn allows them to see much fainter objects. (Fig. 5 compares refracting and reflecting telescopes.)

Seeing faint objects is exactly how William Herschel, the foremost astronomer of the eighteenth century, occupied himself when he received the telescopic baton from Galileo. But unlike Galileo's, Herschel's telescope was an Archimedean lever that moved the Earth *backward*, back to the favored center, launching the second Great Debate. Once again, dust and desire were the astronomer's most reliable fellow travelers, deceiving Herschel as they had Galileo a century before, in a way I'll soon explain.

Using his enormous 1.2 m (4-foot) wide, 12.2 m (40-foot) long reflecting telescope, Herschel plied the night sky from the English countryside. Despite the telescope's massive size, Herschel had only modest goals for it, hoping to discover new stars and perhaps an interloping comet here and there. Instead, he serendipitously doubled the size of the universe.

After discovering a few of Saturn's moons, Herschel glimpsed a small, unknown, disk-like object. Taking a page out of Galileo's playbook, he christened the object the "Georgian star," after England's King George III. It would later be recognized as a planet and was eventually called Uranus. Though it orbits the Sun twice as

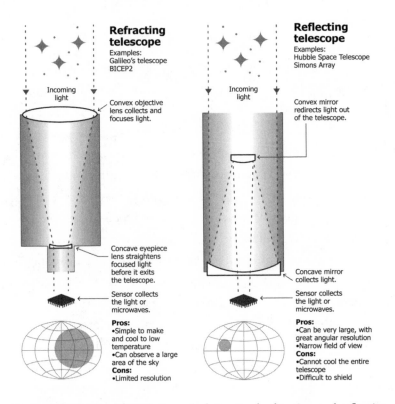

Refracting telescope
Examples:
Galileo's telescope
BICEP2

Incoming light

Convex objective lens collects and focuses light.

Concave eyepiece lens straightens focused light before it exits the telescope.

Sensor collects the light or microwaves.

Pros:
•Simple to make and cool to low temperature
•Can observe a large area of the sky
Cons:
•Limited resolution

Reflecting telescope
Examples:
Hubble Space Telescope
Simons Array

Incoming light

Convex mirror redirects light out of the telescope.

Concave mirror collects light.

Sensor collects the light or microwaves.

Pros:
•Can be very large, with great angular resolution
•Narrow field of view
Cons:
•Cannot cool the entire telescope
•Difficult to shield

FIGURE 5. Comparison of the main features of refracting and reflecting telescopes. (© SHAFFER GRUBB)

far away as Saturn, it is visible to the naked eye, provided you know what you're looking for.

Herschel's leviathan shed light on other dim objects as well. Most important among them were nebulae, the Latin word for "clouds"— which, you'll recall, Galileo believed could be resolved into troves of individual stars. Despite having a telescope vastly more powerful than Galileo's, Herschel could not resolve every nebula into individual stars. Maybe Galileo had been wrong about the true nature of the Milky Way? This question propelled Herschel on the most ambitious astronomical project of the 1700s. After decades of painstaking work, he made the first detailed map of the Milky Way galaxy.

It would have taken longer had Herschel not had a secret weapon:

his genius sister Caroline, who later became the first recognized female scientist in history.[9] In addition to her collaboration with her brother, Caroline Herschel independently discovered many comets, as well as previously unknown nebulae (which, interestingly, were frequently mistaken for comets). She was the first woman to be named an Honorary Member of the Royal Astronomical Society and the only woman to win the society's gold medal, a distinction that lasted until Vera Rubin received it two hundred years later.

When the Herschels' Milky Way survey was done, humanity's ego was restored. The solar system sat smack dab at the middle of the mass of stars (fig. 6). Copernicus be damned: everything outside our solar system *still* was centered on us.

Herschel's chutzpah was forgivable; he was the victim of a powerful illusion. To explain it, I'll use a cloud as a stand-in for a nebula, as well as some physicist's license. An old joke describing a physicist's solution to the problem of a cow that won't produce milk begins with the physicist saying, "Assume a spherical cow. . . ." While I've never seen a spherical cow, as a private pilot I've seen a lot of spherical clouds, or at least they *seem* spherical. Inside a cloud, I can see only a few feet in front of me in all directions, so the cloud appears spherical, like a white shell surrounding me. Fellow pilots compare this visual sensation to being inside a ping-pong ball. The cause of this illusion is a phenomenon called Mie scattering.

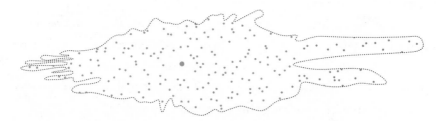

FIGURE 6. Reproduction of Herschel's model of the galaxy. Due to an optical illusion caused by dust, Herschel placed the Sun (gray circle) near the center of the galaxy, far from its actual location more than two-thirds of the distance from the center to the edge. (© SHAFFER GRUBB)

Clouds are made of tiny air molecules and much larger water molecules. Both air and water are transparent, so the white effect isn't caused by the intrinsic color of the molecules. Water droplets are much larger than the wavelength of the sunlight that illuminates the cloud. Light from the Sun may look yellow, but it is actually composed of all the colors of the rainbow, as a simple glance at one of the multicolored arcs after a rainstorm attests. When sunlight scatters off the large water droplets, the colors scramble together and the scattered light appears white. Scattering also attenuates light as it travels to you, meaning that its intensity gets diminished.

Even if the shape of the actual cloud is completely irregular, to a pilot it always appears to be a spherical white shell centered on the cockpit, as in fig. 7.

In 1847, astronomer Wilhelm Struve discovered the reason why Herschel centered our galaxy on the Earth: scattering.[10] The view from Earth is like the view from the cockpit; we are all pilots in this case, aboard a spaceship called Earth. But there the analogy ends, because, instead of water molecules scattering sunlight, tiny grains of dust in the Milky Way galaxy scatter starlight. No matter where Struve looked, the number of stars declined rapidly with distance. He called the combination of attenuation and scattering interstellar absorption, but he didn't know it was dust that caused it. Still, interstellar absorption made it impossible to prove we weren't at the center of the galaxy; even if we weren't at the center, it would still *appear* that we were.

There things stood for another few decades, as astronomers built bigger and better telescopes and finally classified nebulae into two types: those containing individual stars and those seemingly comprised of pure "milkiness," which could not be resolved into individual stars no matter how big the telescope.

In 1912, American astronomer Vesto Slipher returned to Galileo's favorite nebula, the Pleiades, and measured its spectrum. A spectrum is obtained when light passes through a prism or diffraction grating

FIGURE 7. The illusion for pilots is caused by scattering from water; for astronomers aboard spaceship Earth the cause is also scattering, but off interstellar dust, not water. (© SHAFFER GRUBB)

(which you can make using a compact disc). When light is subjected to one of these devices, it separates into its constituent colors. Each color is unique to the atoms that produce it. For example, sodium gas produces yellow light when heated, which is why it is used in many streetlights. Though to the naked eye its light looks somewhat like the Sun's, if you pass it through a prism you will see primarily yellow lines, whereas the Sun's light contains many colors (because the Sun is made of many different elements). For this reason, scientists call a spectrum the "chemical fingerprint" of a light source.

When Slipher spread out the light from the Pleiades nebula, color by color, he found that the nebula's spectrum matched the spectrum of the stars enshrouded by the milky glow. Slipher concluded that the nebulosity was caused by starlight, the light of the same stars scattering off tiny mirrors—dust particles.

This same dusty detritus was what caused interstellar absorption, the illusion that tricked Herschel into thinking we were at the galaxy's center. Galileo and Herschel had both been deceived by the same mirage. Thus, dust clouded the first two Great Debates. It continued to deal cosmic comeuppances to cheeky cosmologists long after Galileo and Herschel.

Soon after the cause of interstellar absorption was discovered, astronomers repeated Herschel's survey of the Milky Way, seeking to correct the illusion he'd fallen victim to and determine Earth's proper place within the galaxy. That would require another astronomical tool.

HENRIETTA LEAVITT RULES
THE UNIVERSE

In 1918, the astronomer Harlow Shapley serendipitously discovered a peculiar type of star within a dense, globe-shaped cluster of stars called a globular cluster. These massive clusters, each comprised of over 100,000 stars, are celestial jewel boxes. They contain within them stellar diamonds in the rough known as Cepheid (pronounced "SEPH-ee-id") variable stars. Shapley immediately realized he'd struck it rich—thanks to the work of American astronomer Henrietta Leavitt.

In 1912, Leavitt announced a remarkable discovery. She found that Cepheids brighten and dim in a repeating pattern, and realized that this predictability could be used to measure the distance to far-off astronomical objects. "A straight line can readily be drawn among each of the two series of points corresponding to maxima and minima," she wrote, "thus showing that there is a simple relation between the brightness of the [Cepheid] variables and their periods."[11]

Using this relation, now known as Leavitt's Law, she determined that the brightness of Cepheids functions as remarkable celestial timepieces. Their periods are inversely related to their brightness; the brighter they are, the slower they pulse on and off. A Cepheid with a one-day period has an intrinsic luminosity (total energy output) of more than one hundred times that of the Sun; a Cepheid with a five-day period is ten times more luminous than that.

Leavitt's "straight line" was actually the longest ruler ever made. Since the brightness of every luminous object decreases with the square of its distance, astronomers can use Cepheids to measure distance. And in 1918, Harlow Shapley did just that. Here's how: when he saw a one-day-period Cepheid at an unknown distance that was one-quarter as bright as a one-day-period Cepheid located ten light years away, he knew that the dimmer Cepheid was twenty light years away, twice as far as the brighter one. It was astonishing: the ticking of a clock became the ticks of a ruler, one that stretched far beyond the solar system.

Shapley used Leavitt's Law to measure the distances to several globular clusters. He plotted their positions and showed they were most certainly *not* centered on our solar system. Instead, they were centered on a more distant point in the Milky Way, toward the constellation Sagittarius, almost a million trillion miles from Earth. Like Galileo with his observations of Jupiter's moons, Shapley used celestial objects that didn't revolve around us to disabuse us of our self-centered tendencies. We are not at the center of the solar system, nor are we at the center of the galaxy. Shapley resolved the second Great Debate in favor of Copernicus: we are far away from the center of the galaxy.

So, we aren't at the center of the Milky Way. But at least the Milky Way is still the entire universe. . . . Or is it? Of course it isn't.

Soon after Shapley discovered the true center of the Milky Way, he began a series of arguments with astronomer Heber Curtis about the existence of other galaxies.[12] These arguments became known as the Great Debate. (Although it was called *the* Great Debate, it is really the third in our series of debates over whether we humans have a privileged vantage point in the cosmos.)

The Curtis–Shapley debate was the first public astronomical debate in centuries. It raged for years, with the two pugilists publishing papers instead of hurling haymakers. Shapley was entranced by the Milky Way's size. It was so big, more than 300,000 light years across, according to his calculations—an estimate we now know to be three times too large—that it *had* to be the whole universe.

How did Shapley, a titan of astronomy, erstwhile defender of the Copernican principle, the man who had just displaced our solar system from the galaxy's center, make such a spectacular overestimate? Once again, the fault lay in the stars, or more precisely in what the stars produce in their dying hour: dust.

DEATH BY TELESCOPE

Galileo's telescope started the series of Great Debates. Herschel's telescope extended the Debates beyond the solar system to the Milky Way's structure, placing us in the privileged central spot. Heber Curtis reasoned that the structure of the galaxy was too irregular for it to be the entire universe. But that wasn't dispositive. If, on the other hand, he could find *another* galaxy, even just one, Curtis could demote the Milky Way to a more modest station and dethrone it from its previous status as the entire universe.

The most conspicuous candidate for another galaxy was the Great Andromeda Nebula. Don't judge a nebula by its name; Andromeda is a galaxy, like our own. But to prove this, Curtis needed an accurate yardstick, and he distrusted Leavitt and her newfangled Cepheids.

To create a different yardstick, he chose novae: volatile stars, known since antiquity, whose intensity flares up on occasion. Curtis compared novae located in the Great Andromeda Nebula to novae

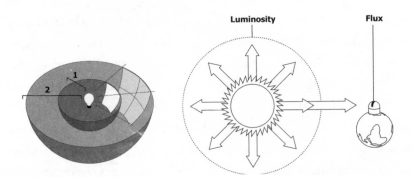

FIGURE 8. Light becomes geometrically dimmer with distance. *Left*: A light source viewed twice as far away appears one quarter as bright because the same luminosity is diluted over a larger area. The illuminated area increases as the square of the distance, decreasing the detected flux by the inverse of the distance squared. Since telescopes detect the flux incident upon them, anything that reduces the observed flux (such as dust in the foreground) can mimic the effects of larger distances between source and observer. (© SHAFFER GRUBB)

in the Milky Way. Those in the Great Andromeda Nebula appeared much dimmer, which meant, to Curtis, either that the novae were very far away and dimmed by the inverse-square law (see fig. 8), or that he had discovered an entirely new class of nova. Conjecturing new phenomena to solve old problems is as unsatisfying to an astronomer as fast food is to a gastronomer. Sure, it does the job, but you hate yourself afterward.

Curtis would win the third of our Great Debates by technical knockout; astronomers generally accepted that the Great Andromeda Nebula must be a great distance away. But its exact distance was not known. Measurement of the extraordinary distance would not come for several years, and it would come from an unlikely source: his rival Harlow Shapley's beloved Cepheids.

THE STAR THAT SHOOK THE COSMOS

October 5, 1923, was just an ordinary night for Edwin Hubble. He was working on the 254 cm (100-inch) diameter telescope on Mount Wilson, north of Pasadena, California. It was an enormous beast, twice as wide as Herschel's leviathan. To guide its ponderous movements, astronomers had to sit in a cage that was attached to the telescope's side. Hubble must have felt like a Maori chief riding a whale, gracefully coaxing the giant beast to plumb the depths of the cosmos.

On that evening, Hubble was photographing the Great Andromeda Nebula. When he developed the film, he saw that he had captured what he thought was an ordinary nova located on its outskirts. He had seen many of these before. He calmly marked the star's position in the photographic plate with the letter "N," for nova, the ordinary one-off flaring star that Curtis had relied on to surmise Andromeda's distant location. Since the brightness variation of a nova doesn't repeat, it was impossible for Curtis, or anyone else, to measure exactly how far away the Great Andromeda Nebula was.

The star Hubble marked turned out to be no nova. Returning the

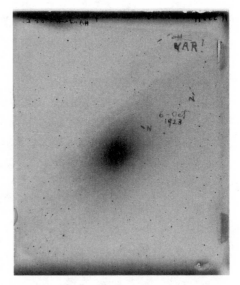

telescope to the same site a few nights later, Hubble was startled to
see that the star's brightness varied periodically; it didn't brighten
once, like normal novae did, flaring up only to rapidly fade away.
Instead, it pulsed in the same way as Leavitt's clock-like Cepheid
variable stars did.

Hubble immediately realized he could use this precious star as a
calibrated ruler, just as Leavitt had done more than a decade before
him. Soon after discovering it, Hubble scratched out the letter "N"
on his photographic plate and replaced it, in big, bold, capital letters, with "VAR!" for variable (fig. 9). Hubble's excitement is pal-
pable when you look at his plate. Using Leavitt's Law, he calculated
that the nebula was more than 2.5 million light years away from
Earth: a tremendous distance. It was located at a distance nearly
ten times greater than the diameter of the Milky Way. The Great
Andromeda Nebula couldn't be in our galaxy. It had to be a galaxy
in its own right. And that meant the Milky Way couldn't be the
entire universe, although it is an admittedly important portion of
it. Another prick to the cosmic ego, another triumph of the Copernican principle.

TODAY, WE KNOW that the Milky Way galaxy (fig. 10) is made of three main parts: a central bulge, a thin disk, and a large, spherical halo. Shapley was right when he claimed that the globular clusters fill the halo. The center houses a massive black hole, a gravitational sinkhole of sorts, around which whirl the Milky Way's stars, gas, and dust.

The thin disk, where the Sun is, is made up mostly of young stars, gas, and dust. Dust can both reflect light (as it does in the Pleiades nebula) and absorb light, as soot rising from a chimney does. The dusty, spiral arms of the disk are where most new star systems form, and it is in the disk that dust absorbs the most light. Shapley hadn't realized that the globular clusters, and the ruler-like Cepheids

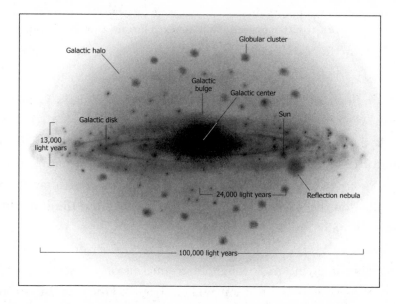

FIGURE 10. The Milky Way is more than 100,000 light years across, with the Sun located in the disk and the center of the galaxy located nearly 24,000 light years away from us. Dust is predominantly found in the disk. Globular clusters—balls of hundreds of thousands of stars—orbit around the galactic center, far above and below the disk, where there is less dust to absorb their light. (© SHAFFER GRUBB)

FIGURE 11. The effect of dust. On the left, a streetlight is viewed through a clean window. If the window becomes dusty, as on the right, the streetlight appears dimmer. If you were unaware of the dust, you might incorrectly assume the light has moved farther away from you. (© SHAFFER GRUBB)

within them, were in relatively dust-free zones, unlike the Cepheids in the disk, which he had used to measure the galaxy's girth. The dark, dusty regions attenuated light from distant sources, making it appear dimmer, just as a streetlight looks dimmer through a dusty window than through a clean one (fig. 11). If you were unaware of the dust on the window, you might be deceived into thinking that someone had moved the streetlight.

Mark Twain is credited with saying that "history does not repeat itself, but it rhymes." Throughout cosmic history, dust has been a part of every astronomical stanza. Dust beguiled the greatest astronomers in history, from Galileo to Herschel to Hubble, and it will continue its divine rhyme in the debates yet to come. From dust we came. To dust we shall return.

Chapter 4

THE BIGGER THE BANG, THE BIGGER THE PROBLEMS

"Every man is trying to either live up to his father's expectations or make up for his father's mistakes."

BARACK OBAMA, *THE AUDACITY OF HOPE*

EDWIN HUBBLE WASN'T USED TO LETTING HIS father down. As an outstanding scholar and athlete, he was a parent's dream; eventually he won a Rhodes Scholarship to study at Oxford University. Although Hubble had been an avid amateur astronomer since boyhood, he abandoned his astronomical aspirations to please his father, who wanted the lad to have a career in which he could make a name for himself. So Edwin dutifully read law at Oxford—but managed to sneak in a few astronomy lectures along the way; the siren song of the stars was too loud to ignore. Soon after his father died in 1913, he quit the legal profession, reportedly saying, "I'd rather be a second-rate astronomer than a first-rate lawyer."

As we have already seen, Edwin Hubble was anything but a second-rate astronomer. By 1923, just five years after earning his PhD (and volunteering to serve in World War I), he'd pulled off the greatest promotion in cosmic history, turning what was considered

a mere nebula into an entire galaxy: our nearest neighbor, Andromeda. But even that titanic feat wasn't grand enough for him.

Hubble proceeded to mine for Cepheids in every galaxy he could
find. The more galaxies in which he could find Cepheids, the larger
the universe became. Using the tools of spectroscopy—the process
of measuring the spectra of astronomical objects which Vesto Slipher perfected—Hubble realized he could measure not only distance, but also the speed at which the galaxies were moving. Slipher
himself had tried to do just this in 1917.[1] Astronomers had already
discovered the phenomena of blueshift and redshift: the spectrum
of an astronomical object appears bluer if it is moving closer or reddened if it is receding. This phenomenon is the optical analog of the
more familiar audible Doppler shift which describes how the pitch
of, say, an ambulance siren behaves: as the ambulance comes closer
the pitch, or frequency of the sound, increases, and as it goes past
you and continues away, the pitch decreases, or lowers in frequency.
Both redshift and blueshift are actual shifts—the colors of light
actually change on the spectrographs, as photographs of the astronomical objects are called.)

Hubble began with Andromeda, and found that our neighbor
was moving toward us, albeit slowly. This made sense, since the two
galaxies have a mutual gravitational attraction. With a handful of
other nearby galaxies, the same convergence was occurring. The
universe seemed to make sense. Until, one day, it didn't.

~

IN 1922, THE YEAR before Hubble found the famous VAR!
star, Einstein received the Nobel Prize. To call his award anticlimactic
would be an understatement; he had even committed the proceeds of
the forthcoming prize to his by-then ex-wife, Mileva, as part of their
1918 divorce settlement.[2] While many people think Einstein won the
prize for creating the special theory of relativity, the award did not,
in fact, cite that accomplishment. Nor did he win his Nobel for what
physicists consider his greatest accomplishment: his general theory

of relativity (GR), which describes how anything with mass bends the fabric of space-time, thus affecting how light travels through the cosmos. No: these two purely theoretical discoveries didn't warrant the Nobel Prize; they represented "world bluffing Jewish physics" in the minds of the men who nominated candidates for the award. Instead, Einstein won his prize for describing aspects of the photoelectric effect, a phenomenon discovered by 1905 Nobel laureate Philipp Lenard. Perhaps the committee's reluctance to cite general relativity as the reason for awarding Einstein the Nobel Prize is understandable; even Einstein himself was uncomfortable with its implications.

Soon after completing it in 1916, Einstein demonstrated that GR could explain phenomena that Newton's law of gravity, the eighteenth-century theory that described gravitational attraction, could not—for example, bizarre features of the planet Mercury's orbit. By the following year, Einstein was emboldened. He wanted to see if the laws of GR held beyond our solar system. After all, if Copernicus's principle was correct, GR should apply to the entire universe, which at that time was still thought to extend no farther than the outskirts of the Milky Way galaxy.

Alas, there was an immediate problem. Einstein found that, according to his GR equations, the universe should be getting *smaller* with time because of the gravitational pull of all the stars. Yet the Milky Way didn't seem to be contracting; stellar spectroscopy showed that distant stars were moving toward *and* away from Earth in equal numbers. So Einstein modified his theory to match the observed properties of the cosmos. As you'll recall, this was long before Hubble's observations; at the time everyone, Einstein included, thought everything beyond the solar system and a few nearby stars was stationary. To make his GR model comport with the static universe, he inserted a cosmic fudge factor, later christened the cosmological constant, into his equations as a type of "antigravity" that explained the universe's embarrassing failure to contract. For the moment, the universe stood still.

In 1922, after Einstein finally received the Nobel Prize, a Russian cosmologist named Alexander Friedmann decided to take another look at Einstein's universe. Friedmann wondered what would happen if he extended GR to include all matter and energy in the universe. Given that there was both matter and energy in the universe, he found, to his astonishment, that the cosmos must either contract *or* expand; it couldn't stand still, as Einstein had asserted in 1917. Contraction made sense because gravity was a known attractor. But the idea of expansion intrigued him: how might that happen?

Sadly, Friedmann wouldn't live long enough to find out. In 1925, on his way back from his honeymoon, he contracted typhoid fever and died, his brilliant life cut short at thirty-seven. But in his brief career, he influenced two superb cosmologists who would carry his torch. One was his PhD student George Gamow; the other was a Belgian Roman Catholic priest and astronomer named Georges Lemaître, who worked at the Catholic University of Leuven.

In 1927, Lemaître explored the implications of Friedmann's model: what if the universe wasn't contracting, but expanding instead?

Einstein mocked Lemaître's idea, calling it an "atrocious" misapplication of his theory. With the paucity of extragalactic observations at the time, his condemnation was not without merit. But by 1927, it was known that many other galaxies lay beyond the Milky Way and Andromeda, and increasingly their spectra showed a bias toward redshift—meaning that more than half were receding from us. There seemed to be some evidence for expansion, but it was hardly convincing; the data Lemaître worked with were horribly imprecise.

Unfortunately for Lemaître, Hubble was on the case, and equipped with the most powerful telescope in the world, that 254 cm (100-inch) instrument on Mount Wilson which he had used to measure Andromeda's almost incomprehensible remoteness.

ON THE SPECTRUM

In 1929, after years of painstaking Cepheid variable hunting and spectroscopy, Hubble boldly postulated the law that is named for

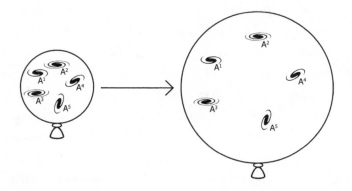

FIGURE 12. In the Lemaître–Hubble model, the universe expands like the surface of a balloon, increasing the distance between all observers. There is no "center" of the universe, just as there is no center of the *surface* of the balloon. (© SHAFFER GRUBB)

him: the farther away galaxies are, the faster they are receding from us.[3] It was a bold assertion; the data at the time were still pretty crude. But like Shapley's claim that the Milky Way was of an enormous size, and Galileo's conjecture that the universe was not centered on the Earth, correct descriptions sometimes come from imperfect theories. Often, even imprecise data firms the footing of a good theory. But, *caveat* cosmologist: usually that's not the case! Eventually, the relation between a galaxy's distance and its redshift/recessional velocity came to be called Hubble's Law. Even though he failed to become a lawyer, Hubble's father would have beamed with pride knowing that his son had his own law.

Hubble's Law implied that the universe was not only dynamic, but that it was expanding like a balloon (fig. 12). Not everyone believed it. It seemed to violate the Copernican principle: suddenly we were special again, if almost everything was receding directly away from us. Imagine being at a big dance party. Attendees around you start to dance. On average, about half the attendees would be moving away from you and half would be moving toward you. Now imagine if you find yourself at a party and *everybody* is moving away from you, you would have to conclude that you were guilty of some sort of social *faux pas*. It was even worse for Hubble's expanding universe. If you took the notion of expansion seriously, madness followed—running the clock backward, the universe would have had a beginning, a special time before which time did not exist. What cheek Hubble had—to take observations of nearby galaxies so seriously! It was most economical to assume that the universe had no beginning, was eternal, and that the recession of galaxies was a mere optical illusion. Eternal universes were so attractive that even Einstein kept studying them years after Hubble's 1929 measurement.[4] Thus began the fourth Great Debate: the question of whether there was both a privileged place and *time* in the universe's history.

Imprecise as it was, cosmologists had their first quantitative data. The only thing that was certain was that the universe was changing. But, just as Galileo's observations didn't prove the heliocentric model but did disprove an Earth-centered solar system, so too did Hubble disprove the static universe without proving the universe had a beginning. But if it did have a beginning, how would we ever glimpse it? And what, if anything, could cause the universe to begin?

In a brilliantly prescient 1931 article in the journal *Nature*, Lemaître presented the first inklings of what we today call the Big Bang. He didn't call it that—you'll find out soon who did. He called it the "primeval atom" model of creation, and it posited that the universe could have begun "in the form of a unique atom, the atomic weight of which is the total mass of the universe." Afterward, the universe would have started its expansion, the one he had originally predicted (and for which Hubble gets most of the credit).

Not only did Lemaître propose the Big Bang, he also proposed some of the first ideas of what we now call quantum gravity, a theory that combines Einstein's GR with quantum mechanics, the laws of physics that describe the behavior of light and subatomic particles like electrons. Prescience came with a price, however. In Lemaître's case, the idea of the quantum origin of the universe was too silly for anyone to take seriously. Quantum mechanics itself was just being fleshed out in the 1930s, and uniting it with gravity remains elusive to this day. So, as the 1930s came to an end, the jury was still out on the fourth Great Debate.

"In all failures, the beginning is certainly the half of the whole."

GEORGE ELIOT

LEMAÎTRE'S MODEL, the one that began with the universe's matter and energy compressed into a single tiny, perhaps infinitesimal speck, upset the millennia-old orthodoxy of an eternal, unchanging cosmos. It clearly implied that everything had been smaller and denser in the past, and that the universe must itself have had a birth at a finite time in the past. As Lemaître's primeval atom model was developed, it created, along with Hubble's observations, four serious problems: some philosophical, others physical.

Problem 1: Space Hubble's Law found a direct proportionality between the distance of a far-off galaxy (measured using Leavitt's Law) and its recessional velocity (measured using Slipher's spectroscopic redshift work). A direct proportionality meant that a galaxy ten million light years away from us is receding *twice* as quickly as a galaxy five million light years away. The relationship can be inverted; by measuring distance and velocity of a galaxy today, we can "rewind" the universe and estimate the time when all galaxies in the universe were "touching." But continuing the logic meant the universe was much smaller in the past—infinitesimally small when the expansion began. Finite stuff compressed into an infinitesimal volume means the cosmic density was unlimited, as was its temperature.

There was no evidence for such a state. It seemed impossible for the universe to have evolved from a state of infinite anything (e.g., density) to a state with finite values for those same physical properties. How could one be sure that the laws of physics themselves would remain unchanged during this once-in-a-universe event? For example, might the constants of nature—such as the speed of light, say—not change amid an infinitely hot and dense background universe? If that happened, all bets were off.

Problem 2: Time Hubble's data were both imprecise and inaccurate: they suggested galaxies were receding seven times faster than the modern-day value. Hubble's speedy galaxies were receding so fast that they must have started their expansionary journey fairly recently. This led Hubble to dramatically underestimate the age of the universe, at a mere two billion years old, exactly seven times younger than today's accepted value. What fooled Hubble? Like Galileo, Herschel, and Shapley, Hubble's overreach was caused, in part, by that ubiquitous scourge of the cosmos: dust. Dust obscured the precious Cepheids he was using, dimming them and making them appear farther away than they were.[5]

Hubble's age estimate implied an all-too-impetuous universe, one that was younger than certain stars within it. Hubble's universe was even younger than the known age of the Earth in the 1930s. Like a stepson who's ashamed to be older than his father's new wife, Hubble's age estimate became an embarrassment for cosmologists.

Problem 3: Matter Lemaître's model couldn't explain how the matter in the universe came about from nothing. In his 1931 *Nature* paper, Lemaître presaged the "universe from nothing" arguments that are in vogue today, invoking a "quantum fluctuation" of the primeval atom as a possible origin mechanism for the cosmos. But then, as now, physicists were suspicious of a universe arising *ex nihilo*. That the universe had a beginning that was reminiscent of the biblical creation narrative was somewhat embarrassing to Lemaître, the Catholic priest. He went out of his way to distance himself from the theological overtones of his model.[6] But they were impossible to ignore.

Problem 4: Principle Worst of all, Lemaître's primeval atom model violated a conjecture astronomers call the perfect cosmological principle, an extension of the Copernican principle past our solar system, out to the whole universe. The perfect cosmological principle simply extends the banality of points in space alone to events in space and time, i.e., space-time. According to Einstein's theory of relativity, space and time are united in a single entity. So, a universe that had a beginning, *the* most special event in history, violated the perfect cosmological principle, which asserted that no point in space, nor any event in time, could claim to be privileged in any way.

British cosmologist Fred Hoyle found the four flaws in Lemaître's primeval atom model so distasteful that in 1949 he inimically rechristened it "the Big Bang," which allegedly was a British euphemism for orgasm. But Hoyle did more than just shame Lemaître's and Hubble's model. He and his colleagues Tommy Gold and Hermann Bondi set to work on a rival theory, one that would solve all four of the Big Bang's embarrassing problems.

> *"Those who do not learn history*
> *are doomed to repeat it."*
>
> GEORGE SANTAYANA

LIVING OUTSIDE LONDON during World War II, Gold, Hoyle, and Bondi had become friends. One night after the war they saw a movie called *Dead of Night*, an early version of that wretched film genre wherein all mysteries are solved courtesy of a dream sequence. In the film, there's a twist: the dreams keep coming day after day, night after night, a chance to repeat life eternally, though only in cycles that the protagonist recalls in macabre *déjà vu*.

To Gold, the movie was a thriller. He wondered if there was a sort of "cosmic repetition" that could be used to craft an alternative to the Big Bang theory. Soon, he showed that an eternal universe continually creating new matter could explain why distant galaxies were all receding from us, as Hubble had observed. In an everlasting universe, certainly there would be no embarrassing age problem, nor would there be violation of the cosmological principle. The new model became known as the Steady State theory. It was an idea that had come and gone and come back again for millennia. But the Steady State was no sequel; it was novel, for it posited that matter was being continuously created, at all times, albeit at a minuscule rate, throughout the cosmos.

The Steady State was the antithesis of the Big Bang. Where the Big Bang had a singular beginning, the Steady State had none. Where the universe was young, comparatively speaking, in the Big Bang model, it was eternal in the Steady State model. And whereas the Big Bang required the entire cosmos to spring forth from a "primeval atom," there were no such inanities in the Steady State.

However, the Steady State was not without its contrivances. It still needed to account for the seeming variation in time. The Steady State theorists allowed the cosmos to change over time, but only by forming tiny amounts of new matter to preserve constant density. Still, its adherents claimed, wasn't that superior to the creation of everything all at once in a state of infinite density? And the rate of matter creation required was tiny; Hoyle poetically described it as merely "one atom per century in a volume equal to the Empire State Building." This amount was undetectable experimentally, immunizing it from observational falsification.

If the Steady State theory were correct, the newly created matter materialized out of empty space and condensed to form stars via nuclear fusion, which in the late 1940s had only recently been understood. The new galaxies would replenish the universe with matter while magically maintaining the average distance between galaxies. It was a bit byzantine, but it was far more believable than creation *ex nihilo*, the Big Bang's explanation for the origin of all matter.

The Steady State theory could even explain the distance–redshift relationship Hubble had observed. And, best of all, it obeyed the Copernican principle, and it sure as hell didn't look like the creation narrative in Genesis 1:1.

An eternal model, cyclically re-creating creation one atom at a time in the dead of night: Gold had found the perfect Hollywood ending—though, admittedly, it required a good deal of artistic license.

In 1948, Gold and Bondi published a qualitative version of the Steady State theory while Hoyle simultaneously published a separate quantitative account that filled in the technical details. The Steady State had solved all four of the Big Bang's fatal flaws. The fourth Great Debate fight had been joined; the Big Bang had a rival. Here were two models, each unpalatable in its own way, and each with strengths and flaws that could not be more different. What would break the tie?

SOME LIKE IT HOT

Were there any signposts on the road back in time, clues that would help cosmologists comprehend cosmic history? In 1948, the year the Steady State theory was postulated, George Gamow and his PhD student Ralph Alpher discovered an unusual cosmic clock, a device more thermometer than chronometer. It was an atomic nucleus called a deuteron, and it became a timepiece for the ages—or, at least, for the period between one second and about twenty minutes after the hypothesized Big Bang, the most scrutinized epoch in cosmological history.

Gamow hypothesized that an understanding of the formation of the very lightest atoms on the periodic table, the same ones you came to know and love in chemistry class, would shed light on the early universe. Any relics left over from the Big Bang would be the lightest ones, the simplest ones made of the fewest ingredients, namely just a handful of protons and neutrons. These light-weights were the only fossils a cosmic archeologist could hope to date. Gamow and Alpher showed that the relative abundance of the chemical elements would provide a sort of time-dependent thermometer, one that was at its most sensitive during the conflagration following the Big Bang.

In 1932 American physicist Harold Urey discovered deuterium, whose prefix "deu" indicates that there are two particles in its nucleus. (The nucleus of the hydrogen atom is a single proton; deuterium is a chemically identical "flavor" of hydrogen, an isotope, with a proton and a neutron in its nucleus.) The deuteron, as the nucleus of deuterium is called, is essentially half a helium nucleus. A bit of culinary innovation suggests a recipe for making that helium nucleus: "take two deuterons and bake at several billion degrees for about a minute." The oven's heat—which consists of particles of light called photons—could slam two deuterons together hard enough to overcome the repulsive electrical force between the two positively charged protons. (The actual process by which helium forms is a bit more complicated than the simple recipe above.) However, the essential step—the coupling of neutron and proton into the recipe's key ingredient, the deuteron—is an uneasy one. The slightest increase in temperature (okay, more than 10 billion degrees Celsius) will shatter the pair apart. So, to make a helium nucleus, the temperature has to be *above* several billion degrees but *lower* than 10 billion degrees. And there's one further complication: the neutron is unstable. If not bound to a proton within about ten minutes, it decays radioactively.

For there to be *any* deuterium left in the universe to eventually make helium meant that the temperature of the universe dropped below this magic 10-billion-degree value within a period short enough that there were still free neutrons around to form the deuterons—under 600 seconds. Thanks to the instability of neutrons, deuterium became a temperature-dependent "clock," a thermochronometer. What would have caused such a rapid cooling from infinite temperatures to this large, but finite temperature? The expansion of the universe. Things cool as they expand—it's

FIGURE 13.
Harold Urey and
Urey Hall at the
University of California
San Diego in 1966,
by Ansel Adams.
(SWEENEY/RUBIN
ANSEL ADAMS
FIAT LUX COLLECTION,
CALIFORNIA MUSEUM
OF PHOTOGRAPHY,
UNIVERSITY OF
CALIFORNIA RIVERSIDE)

a phenomenon you can witness by spraying an aerosol can. When you release the contents, the gas in the can gets less dense and it cools down.

Prior to cooling below this 10-billion-degree temperature mark, any deuterons that could have formed were instantly obliterated by the high-temperature photons whizzing about. Heavier elements like helium simply couldn't form until things cooled off a bit. At that point, there was a (nuclear) building boom—a burst of helium production. But it didn't last long; within twenty minutes of the infinite temperature beginning, it was all over; the universe was too cool to fuse deuterons into helium nuclei and the process shown in fig. 14 came to an end. By then, the die was cast: the amount of *primordial* helium was fixed. It's remarkable to consider—in a time period shorter than an episode of the sitcom *The Big Bang Theory*, all the lightest elements in the universe could be made.

The three sveltest nuclei—deuterons, and hydrogen and helium nuclei—became the ancient artifacts needed to test the Big Bang model.[7] In 1949, Gamow and Alpher predicted there should be one helium nucleus for every twelve hydrogen nuclei plus a tiny bit of

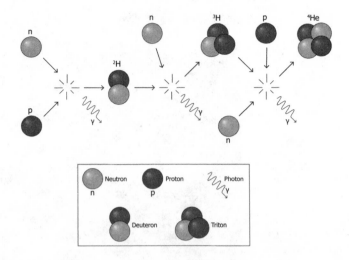

FIGURE 14: How helium was formed in the early universe, according to the Big Bang theory. Unless fused with a proton (making a deuteron), a neutron will decay in about 600 seconds. Luckily, deuteron formation occurs about 100 seconds after the Big Bang, preserving the neutrons. Reactions involving the deuterons and protons and neutrons proceed until helium-4 is produced. The net effect is that deuterons are the key ingredient needed to make helium-4 nuclei. (© SHAFFER GRUBB)

leftover deuterium (which was unobservable at the time). Astronomers found broad agreement with Gamow and Alpher's helium forecast when they observed stars in the Milky Way. Their work emboldened them to suggest how all the elements, even carbon—a heavier element, and the stuff of life—could have been made in the first few minutes of the primordial fireball.

Later, Alpher and his colleague Robert Herman predicted that the universe would keep cooling long after deuterium was forged into helium; in fact, the universe would still be cooling to this day. The heat left over from the fireball, they said, would appear today as a microwave glow in the background. No matter where you looked, you would see the universe glowing at the frosty temperature of 5 kelvin—five degrees Celsius above absolute zero.

That the universe gracefully tapered from a boiling, roiling singularity—a state of matter and energy where anything goes—to a downright frigid bath of light surrounding us in all directions was too ridiculous to take seriously. In fact, only one cosmologist was bold enough to do so: Fred Hoyle. And Hoyle's only interest in it was to continue his Big Bang bashing.

DISAVOWING CREATION

The Alpher–Gamow–Herman theory of the formation of the chemical elements, later called Big Bang nucleosynthesis (BBN), was unpalatable to Hoyle. Still, it incentivized him: surely he could explain the whole shebang, all the matter that matters, in the context of the Steady State model. In doing so, he'd not only kill off the rival theory of cosmogenesis, but also propose revolutionary ideas of his own.

To the Big Bang's four original sins, Hoyle added two more. First, he proved that the BBN model, on the off-chance it was correct, could make only the three lightest chemical elements on the periodic table: hydrogen, helium, and lithium (with three protons in its nucleus), and their isotopes. The Big Bang purportedly began with the universe expanding from an insignificant speck containing the seeds that formed the light elements which, in turn, became the building blocks of the heavier elements. But the Big Bang model didn't allow for the formation of elements heavier than lithium. A theory that could produce only three of the nearly one hundred elements and isotopes known at the time—a model that was only 3% correct—was hardly confidence-inspiring. But Hoyle didn't stop there.

Hoyle's second attack focused on the prediction by Alpher and Herman that leftover heat would still be observable, in the form of a 5 kelvin bath of microwaves. Again, there was a problem. Because things cool as they expand, and the universe was supposed to be expanding over time in the Big Bang model, Hubble's *underestimate* of the age of the universe caused Alpher and Herman (and later Gamow) to *overestimate* the temperature of the bath of microwaves.

The younger the universe, the warmer it would have to be. Alpher and Herman's estimate was, in fact, twice as high as the maximum value of the universe's background temperature could be, according to measurements made in 1941.[8]

More embarrassingly, Alpher and Herman couldn't even make up their minds; the year after they made the 5 kelvin prediction, they presented another value. This time the cosmic temperature jumped to 28 kelvin.[9] Their indecisiveness did little to engender confidence in the Big Bang.[10] In contrast, Hoyle's model had no thermal background whatsoever, let alone a microwave glow that was inconsistent with observational data and varied with each successive publication.

But Hoyle did more than destroy. He also theorized a way in which all the elements, not just the first three, could have been made. They had to be made within stars, he believed. This was a critical hurdle—one that the Big Bang never got over. Recall that the Steady State needed not just an explanation of elements heavier than lithium, but an actual source of new nuclei. Hoyle posited a source as commonplace as could be: stars. Soon he would show how stars could make arguably the most important element for life: carbon.

In 1954, Hoyle showed that to form carbon, a miracle had to happen.[11] Well, he didn't exactly call it that. But based on the fact that you and I are made of carbon, and the fact that there are no nuclei with atomic masses in the right range to form carbon anywhere else in the universe, Hoyle concluded that there must be a special state of matter within stars that somehow catalyzes the formation of helium into carbon. This special catalytic state he called a "resonance," and it's now known as the Hoyle state (fig. 15). Without the Hoyle state, you wouldn't be here reading this.

Three years later, a group led by Caltech's Willy Fowler discovered the physical process.[12] Hoyle was right: only in the guts of stars could carbon be concocted. Miracles do come true.

Success emboldened Hoyle. He had done more than merely refute the Big Bang, more than just pound the sixth, and presumably final,

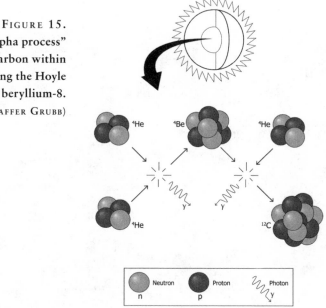

FIGURE 15.
The "triple alpha process"
that forms carbon within
stars, featuring the Hoyle
state of beryllium-8.
(© SHAFFER GRUBB)

nail into its coffin. He had made a bold prediction that could have been proven wrong—but it wasn't.

In 1957, Hoyle began a collaboration that would later yield a Nobel Prize, though not for him. Working with Willy Fowler and Geoff and Margaret Burbidge, Hoyle devised a complex model of the formation of all the elements within stars—known as stellar nucleosynthesis, as opposed to Big Bang nucleosynthesis—that was byzantine in its complexity but amazing in its efficacy. In their epochal 1957 paper, forever known simply as "the BBFH paper" after the initials of its authors' surnames, the quartet showed how all the known elements could be formed within various types of stars. In the end, it came down to Margaret Burbidge, the only true astronomer of the quartet, to produce the hard data that catapulted BBFH into the scientific canon forever (and provided a Nobel Prize for Fowler).

Meanwhile, Gamow and his former graduate student Alpher were still struggling to find evidence for the Big Bang. In fact, it would

FIGURE 16. Margaret Burbidge in action at the University of California San Diego in 1966, by Ansel Adams. (SWEENEY/RUBIN ANSEL ADAMS FIAT LUX COLLECTION, CALIFORNIA MUSEUM OF PHOTOGRAPHY, UNIVERSITY OF CALIFORNIA RIVERSIDE)

take years for observations to catch up to the sophistication of the rival Steady State theory or to match the hard evidence for Hoyle's miracle. Hoyle was merciless. In the final draft of the BBFH paper, he described the formation of the light elements in the rival Big Bang model as requiring "a state of the universe for which we have no evidence." In contrast, stellar nucleosynthesis was a state of the universe for which we had over 100 billion examples in our galaxy alone.

MORE CRACKS EMERGE IN THE BIG BANG

The BBFH paper asserted that all the chemical elements could be made in stars, as long as some were massive enough. In contrast, for the Big Bang, Gamow contended that nuclei had formed long before any stars were born, let alone massive ones.[13] Besides, Gamow asserted, big stars died young; there wasn't enough time for them to produce the heavy elements. But in 1958, astrophysicist Edwin Salpeter showed that, despite their short lifetimes, massive stars could still produce heavier elements, and there were in fact enough of them to do the job. The Steady State passed another test, while the Big Bang flunked again. If you couldn't trust the Big Bang model's explanation for the formation of the heavy elements, why would you trust it for the light elements?

One final point stoked the debate between the competing cosmogenesis camps, and it was not a scientific one. Embedded within the conflict was a religious debate. Many atheist scientists were repulsed by the Big Bang's creationist overtones. According to Hoyle, it was cosmic chutzpah of the worst kind: "The reason why scientists like the 'big bang' is because they are overshadowed by the Book of Genesis." In contrast, the Steady State model was the rightful heir to the Copernican principle. It combined the banality of space with humanity's mediocrity in time. Thanks to Hoyle, humanity had humility. The stakes hadn't been higher since the first Great Debate, which pitted God against Galileo.

For a time, it looked as if the Big Bang was destined for the astronomical ash heap. Then suddenly, in 1964, like a heavyweight champion felled in the final round, the Steady State model suffered two devastating blows.

The first wound was self-inflicted. In a paper published in the fall of 1964 titled "The Mystery of the Cosmic Helium Abundance," Fred Hoyle and his colleague Roger Tayler announced an astonishing discovery.[14] By examining a variety of astrophysical sources, from the Sun to the Orion nebula, they had found that the amount of helium in the cosmos was far too plentiful to have originated solely from within stars. Though Hoyle, Fowler, and the Burbidges had shown in 1957 that stars could make all the known elements and their isotopes, stars couldn't make *enough* of the second most important elemental building block: helium. Stars were like prolific startups that couldn't scale to mass production.

Where once he had mockingly said that his theoretical adversaries Gamow and Alpher had suggested a model "for which there was no evidence," Hoyle himself had now found the evidence, circumstantial as it was. His 1964 paper concluded that either "the universe had a singular origin or is oscillatory." This final statement was tantamount to Steady State suicide; the "singular origin" was a euphemism, as close to "the Big Bang" as Hoyle was willing to utter. The Steady State was on the ropes.

On May 20, 1964, a telescope that would deliver what proved to be the knockout blow began plying the skies over New Jersey. But unlike Galileo's spyglass, this telescope would deal its deadly strike *against* the Copernican principle.

COMMUNICATIONS BREAKDOWN

While the 1960s was a time of upheaval in American society, the decade was a golden age for physics. A surprising "war dividend" was being disbursed, the result of fierce competition between the U.S. and the USSR. Sputnik, the first man-made satellite to orbit

FIGURE 17.
Dr. Robert Wilson (*left*) and
Dr. Arno Penzias with the
6 m (20-foot) diameter Bell
Telephone Laboratories horn
antenna, located at Holm-
del, New Jersey. (REUSED
WITH PERMISSION OF NOKIA
CORPORATION)

Earth, prompted the U.S. government to shower cash on any project with even a hint of potential military advantage. It also marked
a new era of private-sector science funding, with companies like
Kodak, Bell Labs, and IBM playing the role of modern Medici.

The 6 m (20-foot) diameter antenna at Bell Labs—at the time a
private research arm of AT&T—was built as part of the first attempt
at intercontinental telecommunications, not for radio astronomy.
The antenna was to be used in conjunction with a giant metal-foil
balloon as part of Project Echo—a rather evocative name. The idea
behind Project Echo was simple: send radio waves from the Jet Propulsion Laboratory in Pasadena, California, and bounce them off a
high-altitude, reflective helium balloon back down to the antenna at
Bell Labs. The Echo project's success was compromised by the rapid
falloff of radio-wave intensity with distance, making the receipt of
transcontinental signals quite challenging.

The Echo project's technical challenges were significant, but it was the launch of Sputnik that really deflated the balloon program.[15] U.S. politicians feared they were losing the telecommunications race as well as the space race. A transmitter in space solved many of the problems associated with the Echo balloon. Because a transmitter could amplify radio waves, the signals sent back down to Earth were much stronger. And, because it was located in space, a satellite was visible from much greater distances. The U.S.'s response to Sputnik was the Telstar satellite, which was far superior to the Echo balloon. So, the Bell Labs antenna—and its supersensitive detector system— was freed up, and two radio astronomers, Arno Penzias and Robert Wilson, quickly beat Cold War communications swords into cosmo- logical plowshares. The duo repurposed the massive horn antenna into a telescope for a celestial survey that would soon land the sec- ond blow against the Steady State.

LOOKING FOR NOTHING IN ALL THE WRONG PLACES

Both Penzias and Wilson were trained astronomers specializ- ing in the radio and microwave portions of the electromagnetic spectrum—electromagnetic waves with wavelengths between one meter and one millimeter, which corresponds to frequencies between 300 MHz and 300 GHz, respectively. Wilson had even taken a cosmology class from none other than Fred Hoyle during Hoyle's sabbatical spent with Fowler at Caltech. Both Hoyle and the Steady State model had made a deep impression on Wilson. The two radio astronomers began their footsteps into history with a simple goal: they wanted to find nothing. Why seek out nothing? Because in doing so, they could calibrate the antenna for its new purpose, as a telescope.

While visible light ricochets off every charged particle and grain of dust it encounters, radio waves travel almost unimpeded through interstellar space. But, in 1964, only three decades since radio astronomy itself began—with the discovery that the Milky Way galaxy emitted radio waves—little was known about how the Milky Way would "look" at higher frequencies—if, that is, we could "see" microwaves, electromagnetic waves with wavelengths ten to one hundred times shorter than radio waves.

Astronomers wondered whether the Milky Way's microwave emission depended on where you looked. Expectations were that if you looked away from the disk portion of the galaxy, toward what astronomers call "high galactic latitude," there would be less emission, just as looking toward the zenith offers a clearer view of the stars than looking toward the horizon. Radio and microwave astronomers also expected that the galaxy's emission at higher frequencies (shorter wavelengths) would be much weaker than at low frequencies. When looking away from the disk, with the combination of high latitude and high frequency, they expected to see nothing.

Penzias and Wilson knew that even to survey the galaxy's quietest spot, toward one of its poles and far from its disk, they first had to calibrate their instrument. Calibration is essential to all experiments, but especially astronomical ones, where it's impossible to physically "go to the source" of the signal. To calibrate, astronomers compare the readings of their instruments when looking at a source of radiation with a known value. One such known value is zero, i.e., a place from which no microwaves are coming. Ordinarily you wouldn't think that looking for nothing would be a fruitful scientific enterprise, and you'd be right: try getting funding for doing nothing. But surprising things happen when you stare into the void.

Like William Herschel, the astronomer who only hoped to find some new stars with his ultrapowerful telescope and ended up discovering Uranus in the 1700s, Penzias and Wilson had a modest goal for their supersensitive radio telescope: "we were attempting to make sure that we could in fact measure the absence of radiation from the Milky Way, when we in fact found radiation, which was coming, evidently, from beyond the Milky Way."[16] At first, the radiation was an annoyance. They didn't realize they'd hit paydirt.

Radio astronomers speak of the strength of the waves they detect in terms of how hot, in units of kelvin, a perfectly emitting source, a blackbody, would have to be to produce the signals they see. Thus, radio astronomers will describe a source as producing so many kelvin. It's possible for a radio telescope to actually "take the temperature" of astronomical objects, obtaining the exact readings you would find if you went to the object and stuck a thermometer in it.

In 1961, three years before Penzias and Wilson took over, Edward Ohm, a Bell Labs engineer, had used the same 6 m (20-foot) telescope antenna and found that the instrument was, as he thought, too noisy. Now, no radio telescope is perfect; every instrument has some noise so that, even if you were to look at an astronomical object with a temperature of nearly absolute zero, the instrument will indicate the object is slightly warmer. But Ohm found the telescope was *far* too noisy, even after he made a catalogue of all the possible sources of error, something known in experimental physicist parlance as a "systematic error budget." Ohm added up all the contributions he was aware of, sources that he thought could be producing the excess signal that he'd measured, which was 22.2 kelvin with an uncertainty of 2.2 kelvin.[17] (At first glance, perhaps it's the preponderance of "2"s that seems suspicious, but these deuces weren't what was wild about Ohm's claim.)

Ohm's model of the antenna predicted he should see 19.1 kelvin when he looked at the darkest places on the sky. This value was the sum of all the contributions he could think of. But when his measured value came in higher by almost exactly 3 kelvin, showing the number to be 22.2 kelvin, Ohm chalked up the excess to bad luck, then threw out the discrepant data. In doing so, Ohm committed the cardinal sin of data analysis: confirmation bias. Instead of assuming the signal was real, he ruled it out on the basis of his model for what the signal *should* be. A false negative blunder is actually worse than a false positive: aside from the temporary shock you'd experience, would you rather have your doctor fail to diagnose you with cancer when you have it, or tell you you've got it when you're cancer-free?

Ohm's paper gave the false all-clear; there was no need to include the background heat that Alpher and Herman had predicted in his calculations. I still want to grab Ohm by his 1960s standard-issue pencil-thin black tie: How could you do that, Edward? It goes against everything we teach undergraduates in freshman science lab; you can't cherry-pick which of your data you exclude. But Ohm's punishment was far worse than bad marks on a lab report. He became an astronomical schlimazel losing out on cosmology's first Nobel Prize.

ALL THE LIGHT OHM DID NOT SEE

The author Isaac Asimov once said that the archetypical reaction of a good scientist to a puzzling finding is not "Eureka, I have found it!" but incredulity. Often the non-eureka moment is served with fear—fear of being wrong, fear of the shame and guilt resulting from blunders, be they honest mistakes or not.

Penzias and Wilson were courageous. They didn't chalk up their non-zero results to faulty equipment. Instead, they skeptically set out to find out the source of the annoying hiss: was it real, or was it something they'd done wrong?[18]

First, they turned their antenna in the direction of New York City, only about 50 miles away and rotten with radio towers, an obvious potential source of contamination. After all, you wouldn't build an optical telescope on the glittering Las Vegas Strip. Of course, the original purpose of the 6 m (20-foot) dish was for communications, not for radio astronomy, which is why it was located so close to New York City. To remove the hiss, the duo steered it far away from the city. But the signal persisted, whenever and wherever they looked.

Unable to lay the blame on my fellow New Yorkers, they tried to pin it on other dirtbags. They had found the telltale evidence of a family of pigeons, who had made the cozy, sheltered interior of their giant radio funnel into a nest. Birds will be birds, and these happened to have deposited a significant amount of "white dielectric material" inside the horn. After Penzias and Wilson cleaned up the bird poop, they trapped the pigeons and drove them to Philadelphia.[19] But the birds kept returning to their antenna nest—they are known for their homing abilities, after all. "To get rid of them, we finally found the most humane thing was to get a shot gun. . . . It's not something I'm happy about, but that seemed like the only way out of our dilemma," Arno Penzias would recount in self-reproach.[20]

Even after the pigeons left, with a little bang, the noise remained. Only one source could produce such a signal, from all directions, at all times, in all seasons: a cosmological one. They'd need to prove it, though. Surprisingly, confirmation would come from their competition—who hadn't even suited up for battle.

~

BOB DICKE'S FACE FELL. "Boys, we've been scooped!" he exclaimed to his Princeton collaborators. He'd just gotten off the phone with Arno Penzias. Penzias had gotten Dicke's number from a radio astronomer at MIT named Bernard Burke, who had seen a draft of a paper by Jim Peebles, a talented young colleague of Dicke's at Princeton. Peebles's paper improved upon Alpher and Herman's prediction, from a decade earlier, that if hydrogen and helium were formed billions of years ago, the leftover heat would be visible in the form of a bath of microwave radiation, present in every direction you looked. On the phone, Penzias told Dicke about the puzzling buzz of microwaves he and Wilson had found. Almost as a throwaway line, Penzias mentioned that the hiss couldn't be coming from the Milky Way galaxy. Dicke instantly realized the significance.

Well, Dicke thought, he and his team had missed their chance to discover the cosmic microwave background, but at least they could provide an explanation for it. To make the CMB, the only condition Dicke and company required was a universe that had once been in a high temperature state and had three ingredients—protons, electrons, and photons. In this state, the entire universe was a rather simple plasma. Plasmas are sometimes called the "fourth state of matter," in addition to the more familiar gas, liquid, and solid states. You can think of plasma as an even hotter form of gas, one that is made of charged particles like electrons or protons.

The paper Dicke published as a companion to the paper by Penzias and Wilson made copious reference to Bondi and Gold's and Hoyle's 1948 papers on the Steady State model.[21] In fact, the ultimate cause of the high-temperature phase is never identified in Dicke's paper, nor is the phrase "Big Bang" even mentioned. The only cosmic model mentioned in Bell Labs' own press release dated May 23, 1965, was one positing that the "universe is expanding from a high-temperature collapsed state."[22]

Two months later, Dicke and his colleagues showed that the matter in our universe could have been formed from "the previous expansion of a closed universe, oscillating for all time."[23] The conflagration at the end of each cycle erased all traces of the previous universe. A cyclic universe, according to Dicke et al., "relieves us of the necessity of understanding the origin of matter at any finite time in the past . . . the ashes of the previous cycle would have been reprocessed back to the hydrogen required for the stars in the next cycle."[24] Whether the universe was cyclic or not, it was clear now that the *Steady* State model was finished.

AN EYEWITNESS ACCOUNT

What would it have been like to witness the furnace of creation? For twenty minutes following either the Big Bang or the final collapse of the previous cycle, it would have been pretty exciting: the universe was a nuclear fusion reactor, busily churning out most of the helium nuclei it would ever contain. Then, if you could survive the roiling inferno engulfing you, everywhere you looked you would see a charged particle (proton or electron) zipping around, being constantly bombarded by photons as in a cosmic pinball game. After that, it would have been stultifying.

The monotony ended 380,000 years later. At this tender age, the universe had expanded enough to cool down to below the magic temperature of 3000 kelvin, and would have been filled with intense infrared light. To understand how measuring the CMB's temperature helps cosmologists understand the properties of the universe as it was nearly 14 billion years ago (the currently accepted age of the universe), it's instructive to consider an analogy with something more familiar—boiling water. Let's suppose you're Robert Falcon Scott, on your way to the South Pole in 1911. Being an Englishman, you want to make a cup of tea to fortify yourself. To do so, you boil a kettle inside your frozen hut.

Just as the blubber-oil-powered flame begins to boil the water, the hut fills with a thick, foggy mist. You're blinded, unable to see even your hand in front of your face. Your nonvisual senses are heightened, and suddenly you hear the rustling of a penguin outside. Knowing that penguins taste like the ideal combination of chicken and fish, you rush to capture it for your afternoon tea finger sandwiches. While on the hunt, you lose track of time, the flame burns out, and the water in the kettle cools off. The steam vapor condenses into liquid water once it cools below 100°C (212°F). When you return, penguin in hand, you find that the fog has disappeared and you can see around the hut. Curious as to how long it took you to hunt the awkwardly elusive prey, you realize you can precisely estimate how long you were outside the hut from the temperature of the water remaining in the kettle. Knowing the material properties of water, the temperature from which it started cooling (100°C/212°F), and its current temperature, the calculation is easy.

The 3000 kelvin temperature, hydrogen's ionization point, is analogous to water's 100°C (212°F) boiling point: both are precisely known quantities. So are the material properties of hydrogen. Starting from the known ionization point of hydrogen and the measured CMB temperature today, 3 kelvin, we know that the universe has expanded by a factor of one thousand in all directions in the time that has elapsed since it was 380,000 years old. Remember redshift, the reddening of light produced as the light source moves away from you? Well, in the Big Bang model, when the universe expanded one-thousandfold, all wavelengths of light stretched by the same amount. This stretched the wavelength of the infrared light suffusing the uni-

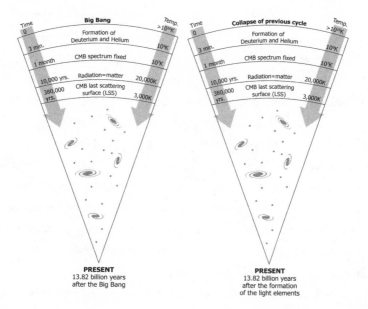

FIGURE 18. Thermal history of the universe according to the Big Bang theory (*left*) and the cyclic cosmological model (*right*). Both theories agree upon the history of the cosmos following the first few minutes after the beginning of time. (© SHAFFER GRUBB)

verse when hydrogen formed (about one micron) to the wavelength of microwaves (millimeters).

When the protons and electrons cooled below 3000 kelvin, the plasma "condensed," not into liquid water but into hydrogen. Hydrogen gas is neutral and transparent to both light waves and microwaves, so after this point (when the universe was 380,000 years old) the whole universe became transparent, just as the condensation of water vapor into liquid water made it possible to see the walls and contents of your hypothetical hut. In the case of the cosmos, the condensation of the plasma into hydrogen enables us to see back in time to when the CMB was formed. This epoch only lasted 100,000 years, a tiny period compared to the 13.8-billion-year age of the universe. In fact, it's so brief a time that we cosmologists think of the epoch when the plasma condensed into hydrogen as a fictitious surface, a 100,000-light-year-"thick" shell known as the "last scattering surface." The last scattering surface surrounds us and represents the shell from which all the photons we now see as the CMB set off on their nearly 14-billion-year-long journey.

In Dicke and company's model, the actual age of the universe was irrelevant. The difference between the actual age of the universe (if the universe had a single beginning, billions of years ago) and 380,000 years later, when hydrogen recombined, is minuscule. And, obviously, 380,000 years is insignificant compared to the "age" of an eternal universe. In both cases, as you went back in time, the universe was hotter and denser, potentially infinitely hot if the Big Bang was correct (fig. 18). It didn't matter who—the Steady Staters or the Big Bangers—was right; there would be a CMB in both models.

Penzias and Wilson were circumspect in their June 1965 paper entitled "Excess antenna temperature at 4060 mega cycles." For what was, prior to BICEP's press conference, called "the most important discovery of all time," one that garnered cosmology its first Nobel Prize, their title really buries the lede. In a valorous example of gratitude and grace, Penzias suggested that Dicke be the third author on the CMB discovery paper. After all, it was only because the two teams were working together that cosmology finally became elevated to the status of a "precision science."

But Dicke demurred, saying he wanted the full credit to go to the two discoverers alone, which effectively removed him from contention for cosmology's first Nobel Prize. Penzias and Wilson received the 1978 prize in physics "for their discovery of cosmic microwave background radiation."[25]

In 1988, the Holmdel antenna site became a National Historic Landmark. It is experimental art, a scientific sculpture as cosmically crucial as Stonehenge. It represents the triumph of generosity over greed, courage over fear, persistence over pride. To me, Holmdel is hallowed ground: mankind's most tangible monument to the quest to comprehend the cosmos.

Chapter 5

BROKEN LENS 1:
The Nobel Prize's
Credit Problem

*"I can forgive Alfred Nobel for having invented
dynamite, but only a fiend in human form
could have invented the Nobel Prize."*

George Bernard Shaw, winner of the
1925 Nobel Prize in Literature

Alfred Nobel was no fiend. He was an idealist,
a dreamer, an inventor who hoped his legacy would be the betterment of mankind. Yet nowadays, his titular prize is causing some
fiendish effects within physics.

This is the first of three chapters discussing what I refer to as the
Nobel Prize's "broken lenses": departures from Alfred's will that
have distorted his vision for world-bettering scientific discoveries.
This chapter focuses on the attribution of credit within science and
the role the Nobel Prize plays as its ultimate reward mechanism.
Future chapters address the role of the Nobel Prize in Physics in
reshaping scientific resources, including human and financial capital. These deviations are distorting the way physicists go about their
craft. Luckily, there are simple reforms that can be enacted immediately, prescriptions that would ultimately help restore Alfred's
noble vision.

OVERDUE DELIVERY

In the Nobel Prize's history, few laureates have waited longer after their discovery than the winners of the 2013 Nobel Prize in Physics. That year, François Englert and Peter Higgs won "for the theoretical discovery of a mechanism that contributes to our understanding of the origin of mass of subatomic particles, and which recently was confirmed through the discovery of the predicted fundamental particle, by the ATLAS and CMS experiments at CERN's Large Hadron Collider,"[1] better known as the Higgs boson.

The discovery of the Higgs boson was announced at the Large Hadron Collider (LHC) in July 2012. A single year between discovery and Nobel Prize doesn't seem like such a long time. It's perfectly consistent with Alfred Nobel's will, isn't it?

Not by a long shot. Englert and Higgs didn't make their theoretical prediction in 2012; they made it in 1964, forty-eight years before winning the prize. No one disputes that they deserved it. But when none of the experimentalists who actually found the "predicted fundamental particle" buried in petabytes of data had a share in the Nobel, more than a few physicists were demoralized. While the LHC's price tag—nearly 10 billion dollars to design and build—seems astronomical, when you understand what the LHC glimpsed in the most sophisticated basement in history, it was a true bargain.

Two different experimental research groups, ATLAS and CMS, took tens of millions of pictures of particle collisions every second. This astounding feat was done more than 91 m (300 feet) below the Earth's surface. The particle collisions took place within a tube 27 km (17 miles) long, which was vacuumed out to a pressure lower than that of outer space. The LHC's twin cameras had to be ultra-precise but robust enough to stop a beam of particles with the energy of a freight train barreling through the tube three trillion times a day, for weeks on end. It took decades to plan and build. It took years to analyze the data. It all culminated in the discovery of the Higgs boson—hardly the overnight success Alfred's will seems to require.

But this was not the troubling aspect of the 2013 Nobel Prize in Physics. What troubled me was that only Englert and Higgs were honored. The committee allows the prize to go to three scientists at most (another "broken lens" which I will discuss in chapter 13). Yet approximately 6,000 scientists worked on the experiments. And even with this arbitrary limitation of three, the prize was awarded to only two physicists, despite there being at least three other living physicists who had played a significant role in the discovery.

One of these overlooked physicists was Gerry Guralnik, one of my mentors at Brown University. His official job was to teach me about advanced topics in quantum mechanics. But I learned much more from him—most of all, what it takes to lead young scientists.

Gerry was a brilliant scientist and a mensch. His paper, coauthored by C. R. Hagen and Tom Kibble, appeared in the same 1964 edition of *Physical Review Letters* as Englert and Robert Brout's paper. Many physicists regard Guralnik, Hagen, and Kibble's paper as at least equal to the contemporaneous papers by Englert and Brout and by Higgs. Of all the papers that could claim priority in predicting the boson, theirs was the one that truly solved the particularly vexing technical issue of the so-called "unwanted Goldstone boson."[2] But Gerry never seemed bitter about losing a share of the 2013 Nobel Prize. After the award to Englert and Higgs, Gerry said, "It is a wonderful feeling of great satisfaction and amazement. We started out to solve an interesting and challenging abstract problem; we were surprised by the answer that turned up."[3]

While he was equanimous, I didn't have to be. I would have used my nomination to right old wrongs and nominated Gerry, but he died in 2014 at the age of seventy-seven, the year before I received my invitation to nominate.

QUEEN OF THE DARK

When I was invited to be a nominator, the first person I thought of was Vera Rubin. Rubin should have been a shoo-in; it was rumored

that she'd been nominated for decades. She had won nearly every other accolade, including election into the National Academy of Sciences—the second female astronomer ever elected to the Academy, after her mentor Margaret Burbidge. Vera's journey into the pantheon of astronomy began at UC San Diego in 1963 when she started a remarkable collaboration with the Burbidges (the two "B"s in the BBFH paper on stellar nucleosynthesis).[4]

Rubin was credited with the serendipitous discovery of dark matter. Like Edwin Hubble, Rubin was transfixed by galaxies and their peculiar dynamics. But whereas Hubble's famous contributions to cosmology focused on the bulk motion of galaxies—nearly all of which moved away from us—Rubin's work concentrated on the odd behavior of galaxies themselves.

Margaret Burbidge had previously shown that the rotation of galaxies was somewhat peculiar, but didn't make much of it. Soon, these cosmic pinwheels became Rubin's trademark tool. Under Margaret's tutelage, Rubin made her first measurement of the rotation of a galaxy, using the methods of spectroscopy. From precise spectra, Rubin measured the rotational speed of stars in distant galaxies. By measuring the slight redshift and blueshift in the frequency of the rotation of stars in the spiral arms in galaxies, Rubin saw something fascinating—the rotation speed didn't die off with distance, as would be expected if stars in galaxies followed the same behavior as planets in our solar system. With her collaborator, Kent Ford, Rubin showed that the failure in falloff was ubiquitous: *no* galaxy behaved like a supersized solar system, which has most of its mass concentrated where its light is densest.

Along her path to the stars, Rubin faced other dark forces. She once considered leaving the field after enduring observatory obstructionism: the lack of women's restrooms on the mountain was the pretext given for forbidding female astronomers to use research telescopes. Rubin countered in a brilliant act of astronomical equal rights assertion. According to cosmologist Neta Bahcall, "She went to her room, she cut up paper into a skirt image, and she stuck it on the little person image on the door of the bathroom. 'There you go; now you have a ladies' room.' "[5]

Margaret and Geoff Burbidge's mentorship was crucial to Rubin's success. "The Burbidges' interest in what I had to say made it seem possible that, yes, I could be an astronomer," Rubin recounted in 2002, showing the vital importance of role models for underrepresented groups in the physical sciences.[6]

There was much controversy over Rubin not receiving the prize in 2016, as there had been for decades. The outcry had intensified after the 2011 Nobel Prize in Physics was awarded for the discovery of dark energy. Many felt the underlying cause was sexism. Others said it was because Rubin hadn't given the interpretation for the effect she and Ford had observed, saying it was "just" a discovery of a phenomenon whose underlying cause was not understood— but, as theoretical physicist Lisa Randall points out, the same could be said about the Nobel-winning discoveries of the CMB, dark energy, and high-temperature superconductivity.[7] Rubin's was painstaking, precision science at its best. Within two decades, she had all but proven the reality of dark matter. As astronomer Jeremiah Ostriker said, "Vera's work, mostly in the early '80s, clinched the case for dark matter for most astronomers."[8] Hubble would have been proud.

The call from Sweden would never come for Vera Rubin. She died on Christmas Day, 2016. Many of my colleagues mourned her passing, not only because of the loss to humanity it represented, but also because so few of them had been alive the last time a woman won the Nobel Prize for Physics. Rubin had seemed like the best shot.

SEASONS OF STOCKHOLM

Like nature itself, the Nobel Prize has its own distinct seasons. Every year, in early October, the season of revelation begins, with the staggered announcements of the six prize winners. That is followed by the season of coronation: the awards ceremony with which we began this book, held annually on December 10, the anniversary of Alfred Nobel's death. There is also a lesser-known Nobel season: the season of nomination, an epoch which closes in the dead of winter, at midnight in Stockholm on January 31 each year. This marks the date by which nominators must submit up to three names. There is no grace period; it is never postponed, and there is no allowance for nominators who tarry.

I expected to use one of my other nominations for the scientists leading the Laser Interferometer Gravitational-Wave Observatory (LIGO) team, which, if the rumors swirling in the astronomy community in late 2015 were to be believed, had detected gravitational waves for the first time in history. These waves, predicted by Einstein's theory of general relativity, reportedly came from two black holes, each more than thirty times the mass of the Sun, caught in a death embrace milliseconds before they merged into one. Like the wake of a boat, gravitational waves are "ripples" in space-time, caused whenever massive objects move. Conceived by Rainer Weiss

and Kip Thorne and brought to fruition by Ron Drever, leading a cast of thousands, LIGO had been looking for shudders in
space-time for decades. I watched impatiently to see if they would
release their long-rumored results before January 31. They didn't.
They refused to rush their analysis just to meet an arbitrary deadline
set by the Royal Swedish Academy of Sciences.

LIGO's detectors had fortuitously come online only weeks before
catching the first gravitational-wave signals on that fateful September day. After months of painstaking analysis by more than 1,000
members of the consortium, the team finally made their announcement on February 11, 2016. Whispers of a Nobel were immediate.
Eight months later, as Nobel revelation season approached, those
whispers intensified. Not everyone was aware that the nomination
deadline had passed before the announcement. Expectations were
that the prize would go to Weiss, Thorne, and Drever—though
many physicists felt that Barry Barish, who had stepped in to
expertly manage the project in 1994, when it was declared "dead in
the water," should be included.[9]

Had LIGO beaten the January 31 deadline, even Barish agreed
that he might have justifiably lost a share of 2016's Nobel Prize due
to the "rule of three."[10] The following year, when the 2017 Nobel
Prize in Physics was announced, the trio was slightly different:
Rainer Weiss, Kip Thorne, and Barry Barish were awarded the prize
"for decisive contributions to the LIGO detector and the observation of gravitational waves." Why not Drever?

While the waves, traveling at the speed of light, took 1.3 billion
years to reach LIGO's detectors, it was a far briefer span of eleven
days that fundamentally altered the 2017 Nobel Prize calculus. Ron
Drever, one of the three founding fathers of LIGO, died at the age
of eighty-five, on March 7, 2016, a year after the LIGO team made
their announcement.

Few doubt the appropriateness of 2017's Nobel Prize winners. Yet
it is impossible to disregard how the Nobel committee solved one
problem—forbidding awards to more than three winners—by vir-

tue of another arbitrary prohibition: one that forbids posthumous prizes. The awful irony of enduring four decades of painstaking work and then dying in the first year Drever was eligible to win science's ultimate accolade seems cruel, almost heartless. And, to add insult to injury, Drever's foundational role in LIGO was downplayed in the announcement of the 2017 prize.[11]

DEAD MEN WIN NO NOBELS

In 1974, the statutes of the Nobel Foundation were revised to stipulate that the Nobel Prize could not be awarded posthumously.[12]

It is ironic that a prize created after the death of its namesake benefactor and two of his brothers is denied to the deceased. Posthumous Nobel Prizes were only deliberately given twice: the 1931 literature prize was awarded to the poet Erik Axel Karlfeldt, and the 1961 peace prize to diplomat Dag Hammarskjöld. I'd be remiss not to note that these men, while surely deserving winners, were both Swedish. Karlfeldt was even the permanent secretary of the Swedish Academy, the organization that selects the Nobel Prize in Literature.

Even after 1974, death could not part two other laureates from their prizes. William Vickrey won the Sveriges Riksbank Prize in Economic Sciences in Memory of Alfred Nobel (formerly called the Nobel Prize in Economics) in 1996 but died days after the announcement. Ralph Steinman was awarded the 2011 Nobel Prize in Physiology or Medicine despite his death days before the announcement. "We are all so touched that our father's many years of hard work are being recognized with a Nobel Prize," said Steinman's daughter, Alexis Steinman, accepting the prize on behalf of his family, adding, "He devoted his life to his work and his family, and he would be truly honored."[13] In a press release, the Nobel Foundation defended their decision to go forward with the award to Steinman despite their revised statute, pointing out that "The Nobel Prize to Ralph Steinman was made in good faith, based on the assumption that the Nobel Laureate was alive" and that what occurred was covered by its "statutes concerning a person who has been named as a Nobel Laureate and has died before the actual Nobel Prize Award Ceremony."[14]

But why go through the contortions in the first place? It would be better to simply allow a posthumous award if a laureate dies within a reasonable amount of time before the awards are announced.

Though some have mocked my suggestion for posthumous awards ("Why not Shakespeare or Homer for literature? Newton for physics? Adam Smith for economics?"),[15] I will counter with an argument based on (Nobel laureate) George Bernard Shaw's famous quip about prostitution. The story goes that, at a party, Shaw told a woman that people would agree to do anything for money, as long as enough money was at stake.

"Not true!" she said.

"Even you would," he replied. "Would you sleep with me for a million pounds?"

"Maybe for that much I would, yes."

"Would you do it for ten shillings?"

"Certainly not!" protested the woman. "What do you take me for? A prostitute?"

"We've established that already," said Shaw. "Now we're just trying to fix your price."

Steinman's death occurred between the decision to award him the prize and the public announcement of it. What if he had died a day before the decision was made: would the committee still have awarded it to him? Possibly. What about six months before? The case of Ron Drever shows that the committee's answer is definitely not. So we just need to fix a date, and it should be fair to all the stakeholders, including nominators. I believe it would be rational to allow the awarding of a posthumous prize if a nominee dies between January 31 and the day the Royal Swedish Academy of Sciences reaches a majority decision on the winner, typically in early October.[16] If that were the case, the input from external subject-matter experts, such as myself, wouldn't be discarded.

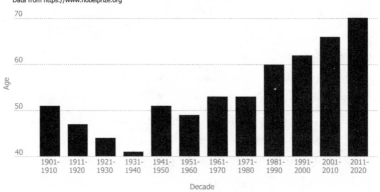

Average age for Nobel Laureates in Physics
Data from https://www.nobelprize.org

FIGURE 20. Average age of Nobel laureates in physics each decade.
(© SHAFFER GRUBB)

But more than honoring the service of nominators, serving science is at stake. Why should the sense of gratitude felt by Ralph Steinman's family not be shared by the next of kin of other recently deceased scientists, even if they died a few months more promptly than Steinman? If anything, it would help to humanize the work of the laureates, and of the Nobel committee as well.

By the time modern-day laureates receive Nobel gold, they are well into their golden years. The average age of physics laureates has increased by twenty-five years since the 1930s, from forty-one to sixty-six (fig. 20).[17] In 2017, the three physics laureates were seventy-seven, eighty-one, and eighty-five years old. The complexity, size, and time-scale of Nobel Prize–worthy experiments are increasing just as quickly, while funding rates decline to levels not seen in decades. One timescale that is rising far more slowly is the human lifespan—meaning that more and more scientists will likely lose Nobel Prizes they deserve.

During the writing of this book, many brilliant physicists, all worthy of the Nobel Prize, have passed away. In addition to Vera Rubin and Ron Drever, Deborah Jin and Mildred Dresselhaus are gone. Failure to award the prize to even one more deserving scientist before death will send a terrible message to physicists: that arbitrary rules supersede objective accomplishments. What will happen to the prize if young scientists feel it is not a true meritocracy, one that recognizes contributions regardless of the scientist's age, political connections, gender, or stature in the field? Awarding a Nobel Prize to one of these deserving but recently deceased scientists would set a wonderful precedent; it would reassure scientists that it is the discovery that matters most, no matter how long it takes.

Chapter 6

ASHES TO ASHES

WHEN I WAS A KID MY FAVORITE COMIC STRIP was *Peanuts*, but my favorite character wasn't the protagonist, Charlie Brown. He was too timid, too neurotic. It wasn't Lucy, either, for she was imperious and cruel. Schroeder was smugly self-satisfied and aloof. The character I adored was Pig Pen, the eternally messy, disheveled kid, enveloped by his own dusty microclimate. He was hardscrabble, nonconformist, salt-and-dust-of-the-Earth. He was ambivalent about social strata, unfazed when kids fled rooms when he entered; his real friends stayed. To him, life wasn't a popularity contest. His dust was his history, and he was proud of it.

Dust complicates cosmology, yet it is the history of the universe, writ small. Dust, ubiquitous and unglamorous, never clamors for attention but cannot be ignored, for dust—star stuff itself—is the very firmament beneath our feet.

Beyond our home planet, could dust play an even bigger role on the cosmic stage, as more than just a sort of cosmic spoiler?

BIG BANG OR BUST

Seven years after the discovery of the CMB, the case against the Steady State model and for the Big Bang model was far from closed.

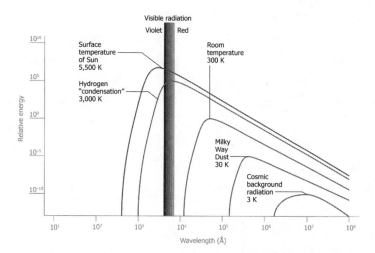

FIGURE 21. A perfectly absorbing and emissive object is called a black-body. The temperature of a blackbody completely determines how much energy it emits at each wavelength of electromagnetic radiation. (© SHAFFER GRUBB)

Even Robert Wilson said that he and Arno Penzias didn't think their discovery of fossil photons spelled doom for the Steady State model: "I guess neither of us took [the Big Bang] cosmology too seriously to start with. In fact, we wanted to leave it open for the steady-state people to propose something to explain this."[1]

In his epochal 1972 cosmology textbook, Nobel laureate Steven Weinberg wrote, "In a sense, this disagreement is a credit to the model; alone among all cosmologies, the steady state model makes such definite predictions that it can be disproved even with the limited observational evidence at our disposal. The steady state model is so attractive that many of its adherents still retain hope that the evidence against it will eventually disappear as observations improve. However, if the cosmic microwave radiation . . . is really blackbody radiation, it will be difficult to doubt that the universe has evolved from a hotter denser early stage."[2] Blackbody radiation means radiation that comes from a perfectly emissive source, one that is at a temperature above absolute zero. Such blackbodies have predictable emission at all wavelengths and their emission spectrum depends only on their temperature (fig. 21). A blackbody is sort of an "anti black hole." Both blackbodies and black holes are completely characterized by their area and one other quality: mass, in the case of black holes, and temperature, in the case of blackbodies. Whereas a black hole *absorbs* all light that lands on it—any color, any polarization state, any energy state—a blackbody *emits* all possible types of light: all frequencies, all polarizations. Anything that gets hot—a glowing poker, an incandescent lightbulb, a ball of hydrogen being fused into helium (a star), anything that emits light or microwaves as it is heated—is a blackbody.

If the CMB's emission could be measured at all wavelengths, the Steady State theory could be disproven—but for most of the three decades after its discovery, no one could do that. Soon after the CMB was discovered, Fred Hoyle and Geoff Burbidge had shown that the total amount of helium in the universe could not have come from stellar nucleosynthesis alone—back-to-back blows for the Steady Statists.[3] Still, the Steady State's principal adherents kept the faith, though they admitted that the theory, in its original incarnation, was unsalvageable.

In their companion paper, published alongside Penzias and Wilson's, Dicke and company explained the CMB within the context of a cyclic universe, one with endless collapse and re-expansion cycles. In the cyclic model, the light elements (hydrogen, helium, and lithium) could arise from "the ashes of the previous cycle," that is, the fiery death of the previous universe could provide the raw ingredients, energy, and matter needed for the next cycle. But from where would the cosmic microwave background (CMB) come, in the cyclic model? And why would it have the exact temperature, 2.7 K, that Penzias and Wilson had just measured?

Hoyle, working with Burbidge, knew the Steady State cosmology had to be seriously retooled. The first thing to go was the model's steadfast steadiness. The name had to morph as well. They rebranded their new model the Quasi Steady State Cosmology, or QSSC. What did the "Quasi" signify?

While the Steady State had originally "begun" without any beginning whatsoever, positing a universe that was eternal, constantly belching out matter and antimatter in just the right proportions to keep the whole Ponzi scheme going, Hoyle and Burbidge's new universe was cyclic, reinventing itself after each oscillation. While it wasn't truly a *Steady* State, the cycles were extremely long: 500 billion years, give or take a few billion. Despite the incomprehensibly long timescales—one hundred times greater than the age of the Sun—this model preserved the Steady State's best features, such as the widely accepted BBFH mechanism by which the heavy elements were made—an explanation which the Big Bang model sorely lacked.[4]

THREE DEGREES OF SEPARATION

Hoyle and Burbidge had already calculated that "the energy released in the synthesis of cosmic helium from hydrogen is almost exactly equal to the energy contained in the cosmic microwave background radiation."[5] Still, just because they could explain the exact amount of energy produced in nucleosynthesis, that didn't mean they could explain why the CMB had the exact temperature Penzias and Wilson recorded. Why was it 2.7 kelvin?[6] Hoyle put that question to the Big Bang model's proponents and they had no answer. Hoyle explained that their predilection for the Big Bang was, like so many before, the result of confirmation bias. He alleged that they had simply done what so many grade-grubbing students do: "plugged and played" to get the "right" answer. "Had observation given 27 Kelvins instead of 2.7 Kelvins for the temperature," he wrote, "then 27 Kelvins would have been entered in the catalogue. Or 0.27 Kelvin. Or anything at all."[7] The CMB's 2.7 kelvin temperature was not a *prediction* inherent in the Big Bang model but a free *parameter* of it, one that required other observations to constrain it.

Hoyle saw the CMB as an opportunity rather than a challenge. He sought to do the Big Bang model one better—to *predict* 2.7 kelvin from first principles. In doing so, he and his colleagues called upon the humblest of all substances: dust.

> *"We come spinning out of nothingness,*
> *scattering stars like dust."*
>
> RUMI

IT'S SAID THAT when all you have is a hammer, you treat everything as a nail. Hoyle's nails were stars. He understood them better than anyone. To bring the retooled Steady State model back, literally from the ashes, and rekindle the fourth Great Debate, Hoyle made use of the coincidence he and Burbidge had found—that the energy made by stars equals the energy in a 2.7 kelvin blackbody. But, as the Sun reveals, starlight isn't in the microwave portion of the electromagnetic spectrum. Stars give off mostly visible light. Therefore, Burbidge and Hoyle had to convert the *visible* light spewed from stars into *microwaves* to accord with Penzias and Wilson's discovery.

If, as Hoyle posited, the CMB had started off not as the relic of the Big Bang but as ordinary, visible starlight, interaction with dust could have transformed it into microwaves. Dust scattering starlight in the nothingness of space, just as Rumi poetically put it.

By the 1960s, it was known that because dust scatters short-wavelength light (such as blue light) much more efficiently than it scatters long-wavelength light (such as red light), the light that makes it to astronomers' telescopes looks redder than it was when it was produced.[8] You can see this phenomenon on a clear night at sunset, when the Sun appears much redder than it does at midday. The Sun, of course, hasn't changed; it's the same 5500 kelvin star all day long. The evening light making it to your eyes appears so red because it is traveling through the greatest amount of atmosphere along the horizon, encountering far more atmospheric dust than it does at high noon (fig. 22).

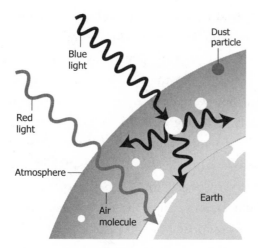

FIGURE 22.
Atmospheric dust and
air molecules scatter
blue light much more
efficiently than red light.
At sunset, the Sun's light
travels through a greater
amount of atmosphere,
which is why the Sun
appears redder at sunset
than at noon.
(© SHAFFER GRUBB)

If the Sun's light is reddened because of a small amount of dust in Earth's atmosphere, certainly the distant starlight from all the stars in the universe would become *extremely* red during its half-trillion-year journey through the dust-ridden cosmos. Indeed, it would be redshifted beyond the visible spectrum to microwave wavelengths. Certainly, scattering of starlight by dust was a much simpler explanation for the CMB than the Big Bang theory, which explained it as the leftover heat from an infinite temperature fireball.

Hoyle knew that dust had clouded even the great Galileo's vision. Now he was claiming that the CMB was just another dusty mirage buoying the hopes of Big Bang believers. But how much dust was there? What was it made of? How was it formed? These questions had to be answered; otherwise, the QSSC model itself would bite the dust.

ANOTHER ONE BITES THE DUST

As a parent of young children, I don't have to look for dust. It finds me. Long before the CMB was discovered, dust was the astronomers' antagonist. It obscured large swaths of the galaxy, causing Herschel to abandon the Copernican principle just over a century after Galileo had proved it. And dust dimmed the light of distant stars, fooling Hubble into thinking distant galaxies were far more distant than they really were. But what is this dust, anyway?

That question vexed astronomers in the first half of the twentieth century. By the 1930s, they were able to estimate the size of the grains causing the redshift.[9] These grains would have to be bigger than the wavelength of red light (about half of a millionth of a meter). That ruled out individual atoms or molecules, which are much smaller. Nor could it be very large chunks, as such stuff would dim all the colors of starlight equally. This left astronomers with only one possibility: dust was comprised of solid particles of a size comparable to the wavelength of visible light. Yet that gave no clue as to the shape and composition of the particles. Were they grains of sand or dirt, such as your vacuum cleaner picks up? Or were they altogether different?

A clue to the makeup of dust came in the 1950s, when astronomers discovered that light from distant stars was not only reddened but also had a new property: polarization.

Electromagnetic waves, such as light or microwaves, oscillate as they travel at the speed of light. This oscillation is what's known as polarization. The direction of light's polarization is always perpendicular to the light ray's trajectory (fig. 23). Polarization provides hints about what happened to the starlight as it made its way to our telescopes. Today, astronomers design devices called polarimeters to detect polarized light, and BICEP2 is one of them. Polarimeters look for differences in the light's intensity within the plane perpendicular to the direction in which the light is traveling.

Not all light is polarized. Most light that comes from heated, glowing objects like incandescent lightbulbs, or the Sun, is unpolarized because the electrons within these objects oscillate in random directions. Unpolarized light, such as sunlight, can become polarized if it reflects off a surface, say, the ocean. The ocean surface is horizontal, so it absorbs light rays polarized *vertically*—that is, perpendicular to the surface. The incoming light that was polarized

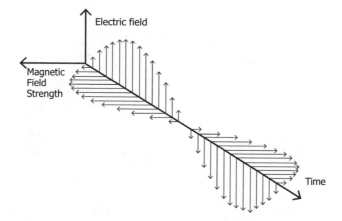

FIGURE 23. The polarization of an electromagnetic wave, like visible light or microwaves, is defined by the direction of the electric field at any moment. The electric field varies in strength and time, but its polarization (in this case vertical) always points perpendicular to the direction the wave is traveling. (© SHAFFER GRUBB)

in the opposite direction—horizontally—ends up being partially reflected. The net effect is that an *unpolarized* beam of sunlight becomes *polarized* after reflection.

Unpolarized light can also become polarized by passing through a filter. Such a device is called a polarizer; it selectively absorbs one of the polarized components and lets the other one propagate unimpeded. Polarized sunglasses make use of polarizers oriented to absorb the light that is reflected from other surfaces—such as the shiny glare from the Sun's light bouncing off the ocean (fig. 24). Removing the glare enables you to see better.

WHAT IS THIS QUINTESSENCE OF DUST?

Astronomers suspected that the same stuff causing starlight to become redder was also polarizing it. By now, the culprit should come as no surprise: it was dust.

Hoyle's model of the QSSC required a lot of dust, not only within

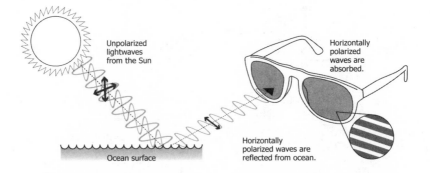

FIGURE 24. Polarizing sunglasses have filters called polarizers that only transmit vertically polarized light. Unpolarized light has equal amounts of vertical and horizontal oscillating light waves. When the unpolarized sunlight reflects from the surface of the ocean it becomes *horizontally* polarized. So your eye, on the other side of the sunglasses, sees no sunlight reflecting off the ocean. This reduces the glare, which is what we call the reflected light. (© SHAFFER GRUBB)

our galaxy, but distributed throughout all space, between *all* the galaxies that existed during a five hundred-billion-year long cycle. And it couldn't be just any dust. In order to transform starlight from the universe's early years into the observed CMB, the dust grains would need to be "whiskers"—cylinders less than a millimeter long. They would have to be made of metal, or at least have some metal grains within them.[11]

Hoyle's metallic whiskers behaved like innumerable little compass needles. They could easily be aligned by magnetic fields, which are ubiquitous throughout the cosmos, banding together in bulk and, like tiny cosmic bouncers, allowing light polarized perpendicular to their orientation to pass through unimpeded while stymieing light polarized along their long axes (fig. 25).[12] The rejected light would bounce around, like drinkers hopping from bar to bar. Eventually, the light would come to thermal equilibrium with a characteristic temperature determined by the shape and composition of the whiskers.

Thus, Hoyle's model not only explained the CMB, it also showed

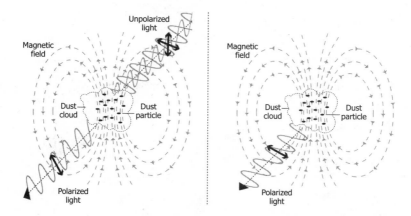

FIGURE 25. Dust in the Milky Way can produce polarized microwaves in two ways. *Left*: unpolarized light from a star passes through a dust cloud. The dust in the cloud is aligned by the Milky Way's magnetic field, which permeates the cloud. The dust particles act like filters, allowing microwaves to pass through if they are polarized along the direction of the magnetic field. *Right*: dust particles heated up by light from distant stars emit microwaves polarized perpendicular to the magnetic field aligning them. (© SHAFFER GRUBB)

why starlight appeared polarized. He was one step away from saving the Copernican principle, killing the Big Bang model, and predicting the precise temperature of the CMB. Thrillingly, he had done it all from first principles, using matter found at ordinary density and temperatures, not at the infinite values the Big Bang needed for both. The only thing missing was the source of the whiskers. What could make such magical whiskers?

Once again, Hoyle would not disappoint. He was a cosmic alchemist and his treasured stars were his unfailing forges.[13] Physicists by now had accepted that fusion within stars made the heavy elements on the periodic table. A star is basically a machine that fuses light elements (like hydrogen) into heavier elements (like helium). When fusion takes place, energy is released in the form of heat. This heat provides the pressure that counteracts the massive force of gravity attempting to implode the star. The process continues as the star fuses helium into heavier nuclei such as carbon, and so on. But once

FIGURE 26.
Anatomy of a Type II
supernova. The star has
lighter elements in its
outer layers when it begins
fusing iron in its core. The
binding energy of iron
does not supply sufficient
heat to keep the outer layers from collapsing. The
core collapse triggers a
runaway explosion, which
sends material out into
the interstellar medium.
Later, these heavy ele-
ments can form the dust
needed to fuel the creation
of new stars, nebulae,
and, eventually, planets.
(© SHAFFER GRUBB)

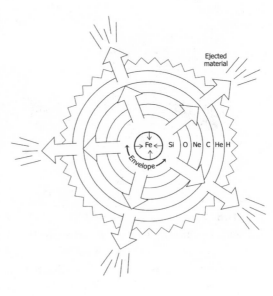

a sufficiently massive star has exhausted all its lighter elements and is fusing iron nuclei together, there is no longer any excess heat energy available to prevent the implosion process. The star collapses, producing a shockwave that blasts the iron out into the surrounding medium, producing what's called a Type II supernova (fig. 26).[14]

Now, the interstellar medium—the stuff between the stars—is a pretty cool and vacuous place. Eventually, the molten iron condenses into solid particles. We have copious evidence for the end product of this process right here on Earth. Every day, our planet plows into tons of micrometeorites, many of which, upon magnification, turn out to be tiny iron whiskers, which scientists believe originated inside supernova explosions (fig. 27).[15]

If the dust particles were iron, as many meteorites are, then they would be magnetic, and the magnetic fields in the Milky Way would

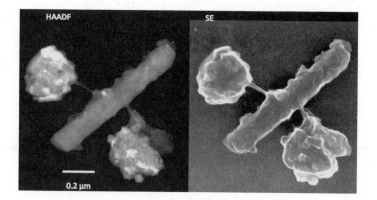

FIGURE 27. Pyroxene (MgSiO$_3$) dust whisker, approximately one-millionth of a meter long, collected from the Earth's upper atmosphere. These whiskers are believed to have been formed in the same cloud of dust in which our solar system was formed. They are made of materials similar to those ejected during Type II supernova explosions. (COURTESY OF DONALD BROWNLEE)

align them. Of course, micrometeorites are not truly interstellar in nature; instead, they are interplanetary media.[16] But that technical detail was overlooked when it was realized that scientists could *make* these dusty metallic whiskers in their labs.[17] When they did, they found that the lab dust had properties remarkably like its cosmic cousin. Best of all, trillions of tons of dust with the properties Hoyle's calculations required were known to exist in almost every galaxy known to astronomers, and much of it was getting blasted out into intergalactic space.[18]

Even after the discovery of the CMB in 1965, most cosmologists doubted the "initial singularity" on which the Big Bang model relied. But no one doubted the existence of dust. No one could claim that supernovae were not real. All the ingredients needed for dust to convert visible starlight into the microwave CMB were in place.[19] Hoyle and company made the necessary detailed calculations and found that their model predicted a background temperature of 2.7 kelvin on the nose.[20] They had done it.

The Steady State theory, in its new incarnation as the QSSC, soared like a cosmic phoenix from the ashes—and it was all thanks to dust, a substance held in contempt by earlier astronomers waging the previous Great Debates. The substance that had caused astronomers to think the Sun was at the center of the Milky Way would deceive astronomers no more. The fourth Great Debate, this one turning on humanity's existence at a propitious time, was resolved in favor of Copernicus yet again. Dust had its day in the Sun.

PER ASPERA AD STOCKHOLM

As soon as the CMB was discovered, in 1965, measuring its emission at all wavelengths became cosmology's holy grail. If experimentalists could do it, they wouldn't prove the Big Bang theorists right—but, almost as satisfyingly, they might prove the Steady State theorists wrong. After nearly two decades of uncertainty, in 1990, the Far InfraRed Absolute Spectrophotometer (FIRAS) instrument aboard the Cosmic Background Explorer (COBE) satellite definitively measured the CMB's "rainbow" (how bright its light was over all frequencies). It matched the predictions for radiation from a blackbody, providing incontrovertible evidence that the CMB was a blackbody.[21] But in the meantime, the QSSC theorists had found a way around this; if the blackbody radiation was coming from starlight, dust whiskers would scatter the light sufficiently to diminish the radiation to the exact temperature FIRAS observed.

Nevertheless, to most cosmologists, FIRAS proved the Big Bang. More Nobel Prizes followed—but not to the theorists who had predicted it. Gamow, the so-called "father of the Big Bang," died long before FIRAS's definitive measurement, only three years after the CMB was discovered.[22] Ralph Alpher didn't win one either—even though he was alive in 2006 when half of the Nobel Prize in Physics went to John Mather, the leader of the FIRAS team, for the discovery of the CMB's blackbody spectrum. (The other half went to George Smoot, the team leader of another instrument on COBE called the Differential Microwave Radiometer, or DMR, which measured the CMB's minuscule degree of anisotropy—deviation from isotropy, or perfect uniformity; it turned out the CMB doesn't have a perfectly uniform glow after all.) Alpher died the following year. Robert Herman died in 1997, which, thanks to the 1974 posthumous stipulation of the Nobel Foundation, means that all the theorists who conjectured the Big Bang are forever disqualified from winning the Nobel Prize.

The Big Bang's opponents, the Steady Statists, also failed to win the Nobel Prize, though Fred Hoyle was awarded a valuable consolation prize in 1997, when the Royal Swedish Academy of Sciences, the same organization that gives out the Nobels, named him the winner of the Crafoord Prize, which came with a purse three-quarters the size of a Nobel. This was seen by some as the Academy's admission of guilt for not giving Hoyle the same credit they gave his collaborator Fowler in 1983. In its Crafoord citation, the Academy commended Hoyle's ingenuity: "In Hoyle's most well-known work on the origin of the elements, a number of the most important stellar processes which have produced the elements were given for the stellar origin of the elements, except for the lightest of them. His later work has in particular been directed towards cosmology and the nature of the interstellar dust. This work is characterised by new interesting ideas and, sometimes, speculation."[23]

The word "speculation" hinted at the QSSC model, and "dust" referred to Hoyle's magical whiskers. Obviously, as late as 1997, the Nobel committee still considered Hoyle's dusty mechanism a plausible explanation for the origin of the CMB, and why it had the 2.7 kelvin temperature it did.

While there were flaws found in the QSSC model, none had yet proved fatal. The QSSC's creators found ways to evade all criticism; they were even able to explain the existence of the tiny fluctuations in the CMB's intensity that the COBE satellite's DMR instrument had discovered in 1992.[24] But explaining the DMR result, as impressive as it was, turned out to be the last victory the QSSC would enjoy.

LOSING BY A WHISKER

Fred Hoyle was spared the agony of witnessing the death of his beloved QSSC. He passed away in 2001, only a year before the QSSC model suffered a self-inflicted mortal wound, courtesy of what was once called its most appealing feature: its falsifiability.

For the QSSC model to be viable required the cosmos to be filled with some very special dust. This dust had many jobs. First, it needed to absorb starlight—which, as we know from the Sun, is almost all in the form of visible-wavelength light. Then, the dust needed to thermalize, meaning that all the dust between the stars had to come into equilibrium at the temperature Penzias and Wilson, and later FIRAS, had observed, 2.7 kelvin. Lastly, in order to equilibrate and not continue heating forever, the dust had to re-radiate energy in the form of microwaves, which would then travel throughout the universe. To be such proficient multitaskers, these dust grains needed to be whiskers, tiny cylinders under a millimeter long.

If the QSSC model was correct, the CMB should be polarized. In fact, after passing through innumerable dust whiskers, the CMB should be highly polarized. Finally, here was a falsifiable prediction made by a cosmological creation story! If the polarization of the CMB could be measured, Hoyle's model could be ruled out.

In 2002, the Degree Angular Scale Interferometer (DASI) made the first detection of the cosmic microwave background's polarization.[25] DASI, led by one of the preeminent experimental cosmologists of our time, John Carlstrom, along with his graduate student John Kovac, definitively showed that the CMB's polarization was nowhere near the amount predicted by the QSSC. It was minuscule, only 0.00001%—a signal about one thousand times lower than Hoyle's whiskers should have produced.[26] The dust wasn't there.

The new polarization data fit nicely into the Big Bang narrative. In the Big Bang model, the CMB was initially completely unpolarized, as is all blackbody radiation. But during the universe's first 380,000 years, its primordial photons encountered many electrons in the primordial plasma, before the electrons clung to protons to produce hydrogen. When these photons hit an electron, the Big Bang model hypothesized that they would bounce off in a process called Thomson scattering. (Physicists name many processes after the scientists who discovered them; this one was named after J. J. Thomson, winner of the 1906 Nobel Prize in Physics.) Unlike the polarization of microwaves by stardust, Thomson scattering produces much smaller amounts of microwave polarization. And that was exactly the measurement that DASI obtained. Other experiments soon confirmed DASI's results, including CAPMAP, a Princeton polarimeter led by Suzanne Staggs, and an upgraded version of BOOMERanG called B2K.

Cosmic microwave background polarization data dealt the coup de grace to the Steady State that neither Hubble, nor helium, nor Penzias and Wilson's discovery of the CMB was able to deliver. Masterful as the DASI results were, they weren't enough to convince the sole survivor of the QSSC tontine: Geoff Burbidge. He continued to extol the virtues of the QSSC model as late as 2009, when he published *Facts and Speculations in Cosmology*, coauthored with Hoyle's former graduate student Jayant Narlikar, with no mention of DASI's factual blow to the QSSC speculation.[27] Nevertheless, after the first measurements of the CMB's polarization, the QSSC theory itself reached the ultimate steady state: it was officially dead.

THE SUN SETS ON THE STEADY STATE

After defending his PhD thesis, John Kovac applied to be a postdoc in Andrew Lange's group at Caltech. He wanted to work on BICEP, which had just received funding from the National Science Foundation. Andrew asked me, as one of the co-leaders of BICEP, to give an opinion on whether Kovac would be a good addition to the lab. It was a simple answer: yes, of course we should hire him. His thesis was magnificent, plus he had local knowledge of the South Pole, where he had spent an entire year of his young life running a CMB

experiment almost singlehandedly. With Kovac on board, BICEP was poised to look farther back in time than anyone ever had.

In January 2010, Geoff Burbidge passed away at age eighty-four. A year later, I was offered the chance to move into his old office at UC San Diego. Although I knew about his illustrious past, it wasn't until I carefully packed up his file cabinets that I realized how titanic his contributions to astronomy really were. He had authored more than four hundred peer-reviewed papers, including some of the most important in all of astronomy—on the nature of the most energetic galaxies, the formation of the elements, and the rotation of spiral galaxies—work Vera Rubin would later use to shed light on dark matter.

For much of the time I knew Geoff, I thought him a cosmic curiosity, quixotically tilting at interstellar windmills. In fact, he was a heavyweight boxer with no one left to fight. Lumberjacks say you don't know how big a tree is until it falls. The same is true of cosmologists and the models they make.

Dust couldn't save the QSSC. But a few years later, on what

would be the fiftieth anniversary of the discovery of the CMB, dust would cloud cosmology again, this time masquerading as another celestial background. Lying in wait for unconscientious cosmologists, the dust abides. In astronomy, as in life itself, it really pays to be fastidious.

Chapter 7

THE SPARK THAT IGNITED THE BIG BANG

*"No one ever made a decision because
of a number. They need a story."*

DANIEL KAHNEMAN, WINNER OF THE 2002 SVERIGES
RIKSBANK PRIZE IN ECONOMIC SCIENCES
IN MEMORY OF ALFRED NOBEL

MUCH AS IT PAINS COSMOLOGISTS TO ADMIT it, the most widely believed theory of cosmogenesis is the biblical creation story. It took a few millennia for the numbers to catch up with the story, in the mid-1600s, when Archbishop James Ussher dated the origin of the universe—Genesis itself—to the night of Saturday, October 22, 4004 BCE. Perhaps Ussher overinterpreted his data: the first two chapters of the Bible.

Today, astronomers wielding advanced versions of Galileo's telescope, like BICEP2, still strive to quantify how and when it all began. It is no surprise why: humanity has always longed to understand its own origins.

The Big Bang model outlasted its rivals, emerging victorious but hardly unscathed. It still had some very large problems to overcome, foremost among them being: what banged? What caused the infinite temperature singularity at the very beginning? The model seemed to violate Newton's third law, a law that every freshman physics student learns: every action has an equal and opposite reaction. The Big Bang seemed to be all reaction, with no action preceding it. And, as we've seen, it was too much like Genesis 1:1 for most practicing cosmologists' tastes.

As telescopes and detectors became more sensitive, astronomers could tease out finer and finer details about the cosmos. Closer inspection showed further cracks in the Big Bang model. For a while, it seemed as if it, too, might not survive.

The Smoothness Problem The strongest evidence for the primordial conflagration, the CMB, continued to beguile astronomers; it was too good to be true. The Big Bang model posited that the CMB originated from a plasma, a gas of electrons and protons too hot to condense into hydrogen, which endured for the universe's first 380,000 years. Plasmas are opaque—no light gets through them—and they are highly uniform. Penzias and Wilson found this to be true of the CMB as well, writing, "This excess temperature [i.e., the CMB] is, within the limits of our observations, isotropic, unpolarized, and free from seasonal variations."[1] The early universe was a hotbed of homogeneity, altogether remarkably boring. For three more decades, no one saw anything that would suggest otherwise.

Shouldn't that be a good thing? After all, the Big Bang *predicted* that the CMB was born out of a plasma, and plasmas are pretty much the same everywhere. Wasn't the CMB's uniformity more evidence for the Big Bang? Not quite. The isotropy of the CMB presented a challenge: if the universe was perfectly uniform, how did stars, galaxies, and planets choose the locations where they formed? If the primordial plasma was a sea of sameness, with nothing to distinguish one place from another, it would stay that way forever. Nothing would cause hydrogen to condense into stars in any particular region of the universe. The paradox was clear—something had to have upset the uniformity of the early universe, otherwise we would not be here to study it.

Finding no evidence for anisotropy—deviation from perfect

FIGURE 29.
The CMB fluctuation pattern is unbelievably uniform, varying by only 0.01% over the whole sky. Reconciling the large, highly inhomogeneous structures we see today, such as galaxies, with the smoothness of the early universe is known as the Big Bang theory's "smoothness problem."
(© SHAFFER GRUBB)

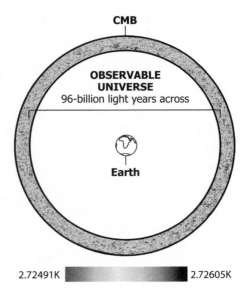

CMB

OBSERVABLE UNIVERSE
96-billion light years across

Earth

2.72491K 2.72605K

homogeneity—in their original data, Penzias and Wilson moved on, content with having made one spectacular discovery. It was another twenty-seven years before the Cosmic Background Explorer (COBE) team announced in 1992 that they had finally detected the all-important anisotropy in the CMB, a discovery described by George Smoot as tantamount to "seeing the face of God."[2] The deviations from perfect smoothness were extremely tiny, however; the CMB's temperature varied by a mere thirty *millionths* of a kelvin across the sky, a variation level 100,000 times smaller than the 2.7 kelvin CMB temperature (fig. 29).

As nearly every new discovery does, COBE's findings created more questions than answers. First, where did the fluctuations come from?

Cosmologists Rainer Sachs and Arthur Wolfe had in fact answered that question in 1967, just two years after Penzias and Wilson discovered the CMB. When light travels near mass, gravity pulls on that light and causes redshift: the shift to redder color, longer wavelength. The longer the photon's wavelength—the redder the light is—the less energy it has. Sachs and Wolfe suggested that, in the early universe, regions with more mass—called high-density regions—gravitationally "tugged" on the CMB photons harder than low-density regions did. The pull is called gravitational redshift. Thus, the slightly cooler regions COBE saw in 1992 corresponded to places in the primordial plasma where matter density, and thus gravity, was ever so slightly stronger than average—wrinkles in space-time itself. These would soon wrinkle the brows of cosmologists.

Though Sachs and Wolfe had explained how variations in mass in the early universe could cause fluctuations in the temperature of the CMB, there was no explanation for how those first mass fluctuations arose, let alone why those fluctuations were small but not perfectly zero. The small, but non-zero, level of fluctuation required became known as the Big Bang's *smoothness problem*.

The Horizon Problem The CMB's temperature was nearly identical everywhere—not only for neighboring patches of the sky, but even for two regions on opposite sides of the sky. These regions corresponded to parts of the universe that were separated by distances far greater than the distance light could have traveled since the Big Bang (fig. 30). Far-flung regions of the cosmos had the same temperature, despite being so distant that they'd never been close enough to one another for their temperatures to equilibrate.

This troubling observation became known as the *horizon problem*. I know what you're thinking: "horizon" implies "edge" which maybe implies there's a "center" of the universe . . . and we are at it! But here, by "horizon," cosmologists mean the maximum distance two events can be from each other such that light has just enough time to reach us from both events.

To understand how problematic this really is, consider a fictitious appliance store that never closes. It sells only one strange model of toaster-oven, one with a fixed temperature of 500 degrees. What it lacks in diversity, though, the store makes up for in quantity: suppose it has in stock thousands of such toaster-ovens and each one has its own sales clerk. The clerks' job is very straightforward; they

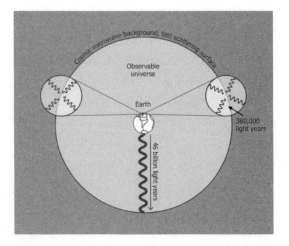

FIGURE 30. The cosmic microwave background comes from 46 billion light years away. However, when the light was emitted the universe was much younger (380,000 years old). In that time light would have only reached as far as the smaller circles. The two points indicated on the diagram were too distant to influence each other because their spheres of causality do not overlap. The fact that such distant regions have nearly identical properties is known as the Big Bang theory's "horizon problem." (© SHAFFER GRUBB)

can only turn the toaster-ovens on or off, and they are free to do so whenever they want.

Ovens that have been on for a while will be at 500 degrees. Ovens that have been off for a while will be somewhere between room temperature and 500 degrees, depending on how long a particular oven has been off.

When you enter the store, what would you expect to see? Probably you'd expect a random distribution of temperatures between room temperature and 500 degrees. Imagine your surprise when you find that all the ovens have the same temperature, say 272 degrees, to within 0.005 of a degree! How can you explain such a finding? It clearly can't be random. It seems as if a conspiracy is at work. But even so, it would be a heroic one: the clerks would have had to synchronize both the exact duration for which each oven is on and the precise time each oven was turned off. The conspiracy requires near-instantaneous collusion throughout the vast store.

In the Big Bang theory, the conspiratorial explanation was to assert that the entire universe was created with near-perfect homogeneity and it remained so forever more, even as it expanded to more than 90 billion light years across. This was unsatisfying, a "just so story" if ever there was one, as much anathema to cosmologists as Genesis.

The Flatness Problem If you've ever looked closely at the surface of a horse's saddle or a beach ball, you may have noticed that it has two primary properties: 1) how "bendy" it is at large scales, and 2) how "wrinkly" it is at small scales. Bending and wrinkling are two types of what mathematicians call curvature.

By convention, a ball is said to have a positive radius of curvature; you might say it is bent outward. In contrast, a saddle has a negative radius of curvature; it's bent inward. And a flat surface, like a sheet of paper, has an infinite radius of curvature, because, well, it doesn't curve. As mentioned above, in 1992 COBE detected that the CMB exhibited variations in the amount of microwave energy from point to point. These fluctuations were caused by variations in the amount of matter/energy from place to place in the primordial plasma, producing wrinkles in the curvature of space-time itself. Space-time was crinkled. But, in 1992, it wasn't known if the universe also "bent" on a large scale. And if it did bend, was its radius of curvature positive (like that of a ball), negative (like that of a saddle), or infinite (like a sheet)?

Measuring the curvature of a surface simply requires making triangles, just as it did for Euclid two millennia ago. Euclid showed that if you drew a triangle on a flat piece of papyrus and added up the interior angles within it, the total would be 180 degrees. But on a curved surface, like that of the Earth, you can make a triangle

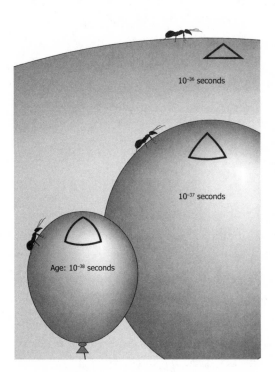

FIGURE 31.
"Ant-flation": An ant crawling over the two-dimensional surface of an expanding balloon illustrates how the theory of cosmic inflation might explain why our universe appears flat on its largest scales. If the ant were to measure the sum of the interior angles of the tri-angle in front of him as the universe (balloon) expands, he would find the sum decreases rapidly and then converges to 180 degrees for all time, irrespective of how curved the universe was when it began expand-ing. (© SHAFFER GRUBB)

10^{-36} seconds

10^{-37} seconds

Age: 10^{-38} seconds

in which the three angles add up to a much larger sum. Imagine a journey beginning on the equator in Quito, Ecuador. Leaving Ecuador, you go due east to Kuala Lumpur, Malaysia, on the opposite side of the planet. Then you travel due south, to the South Pole. Finally, from the Pole, you head due north back to Quito. The angles inside that triangle add up to 360 degrees—twice the sum of the angles of a triangle on a flat surface. Using a simple triangle, you have measured the curvature of a two-dimensional surface. For years, astronomers tried the same measurements using planets, stars, and even whole galaxies. Each time they found no evidence for curvature whatsoever (fig. 31).

Those triangles were relatively small, though, by cosmic standards at least. To measure the curvature of the whole universe requires a triangle that has at least one leg extending as far away from Earth as possible, which means that it ends at a spot that existed right after

FIGURE 32.
The bottom images show the CMB's temperature anisotropy for three different possible universes, with different values of spatial curvature. BOOMERanG showed we live in a flat universe, corresponding to an infinite radius of curvature.
(© SHAFFER GRUBB)

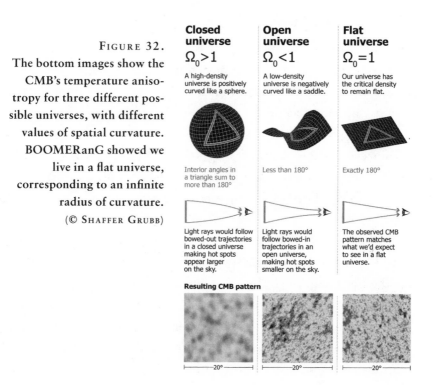

Closed universe
$\Omega_0 > 1$

A high-density universe is positively curved like a sphere.

Interior angles in a triangle sum to more than 180°

Light rays would follow bowed-out trajectories in a closed universe making hot spots appear larger on the sky.

Open universe
$\Omega_0 < 1$

A low-density universe is negatively curved like a saddle.

Less than 180°

Light rays would follow bowed-in trajectories in an open universe, making hot spots smaller on the sky.

Flat universe
$\Omega_0 = 1$

Our universe has the critical density to remain flat.

Exactly 180°

The observed CMB pattern matches what we'd expect to see in a flat universe.

Resulting CMB pattern

|— 20° —| |— 20° —| |— 20° —|

the Big Bang. If such a distant leg could be found *and* that leg had a known length, it could act as a cosmic ruler, allowing the entire universe's radius of curvature to be measured.

It turned out that these "standard rulers" did exist; they were small-scale oscillations—sound waves—that propagated through the plasma that produced the CMB and had been transformed into temperature variations that were about a degree across.[3] They were the biggest structures that could have formed in the adolescent universe, rulers equal in length to the age of the universe multiplied by the speed of sound. The rulers were equal in size to the diameter of the small circles in fig. 30. Eight years after COBE, these fluctuations were discovered by the BOOMERanG experiment and were immediately used to measure the curvature of the universe. It turned out that the radius of curvature was infinite: the universe is flat. No matter how big a triangle you make, its angles always add up to 180 degrees (fig. 32).

Like COBE's, BOOMERanG's findings were revelatory. Our universe really shouldn't be flat. Flatness is unstable, meaning that unless the universe is *perfectly* flat from its beginning, any tiny curvature will become amplified. The observed flatness implied that, at just one nanosecond after the Big Bang, the universe had the so-called "critical density"—where the amount of matter on average within every cubic centimeter is 447,225,917,218,507,401,284,016 grams.[4] If the universe had one more gram per cubic centimeter, out of 447 sextillion, it would be too dense to avoid collapsing in on itself by now.

BOOMERanG's observations indicated that the curvature of the universe had, essentially, always been flat. This presented a crisis for the Big Bang. According to Lemaître, the Big Bang began as a "primeval atom," a tiny speck, with a radius of curvature almost equal to zero—as far as is imaginable from the infinite radius of curvature Lange and his collaborators had measured. Once again, our universe seemed to be born lucky, in a fortuitously finely-tuned state that was flat on the largest scale and nearly perfectly smooth on the smallest scale. Cosmologists hate luck. This inexplicable fortunate flatness became known as the *flatness problem*, the third of an increasing number of seemingly insurmountable problems for the Big Bang theory.

A WILD GUTH CHASE

How did the universe get to be ironed out so Florida-flat, on both large and small scales? It was almost as if the universe had burst forth from the primordial singularity and then expanded so much that all traces of its initial wrinkles were ironed out. Putting it that way almost suggests the answer. Although the actual smoothness and curvature weren't measured until 1992 and 2000, respectively, for decades cosmologists had anticipated the eventual evidence.

In 1978, the year Penzias and Wilson would win the first physics Nobel Prize for cosmology, a young, soon to be unemployed physicist named Alan Guth was working as a postdoc at Cornell University. One day, Bob Dicke, the Princeton physicist who'd led the companion paper purporting to explain the CMB (and voluntarily removed himself from Nobel Prize contention), was at Cornell lecturing on the Big Bang's flatness and horizon problems.

Guth was confused by what he heard. "Dicke's lecture seemed to point to the conclusion that traditional big-bang theory was leaving out something important," he would later say.[5] Dicke, along with his colleague Jim Peebles, argued that the universe must be very close to flat and smooth on anthropic grounds. The anthropic principle is a rather banal-sounding statement: the universe can't be too different from what we observe it to be given that we are here to observe it. If it were too wrinkled, gravity would cause all matter to collapse until nothing was left save some black holes; nary a cosmologist extant to worry about the lack of smoothness.[6] Conversely, were it too smooth, no stars or planets would have formed. Once again, there would be no one to wonder why it was so smooth. But anthropic arguments are as unsatisfying as being a vegan at a Texas barbecue; sure, there's something there, but it's not very filling. Alan Guth hungered for more.

After hearing Dicke's talk, Guth moved from Cornell to Stanford's Linear Accelerator Center (SLAC). It wasn't a permanent job, but it would have to do—he had a young family to support and needed to make his mark to land a faculty job. The Big Bang's failings didn't seem like the stuff of untold riches. But intellectually, at least, they turned out to be just that.

Guth couldn't stop ruminating on Dicke's lecture. One night in December 1979, he scrawled a cryptic note in his notebook. Like Hubble, who, upon glimpsing a treasured Cepheid variable star in the Andromeda galaxy, inscribed "VAR!" on the photographic plate that revealed it, Guth was so thunderstruck his words came out in all caps: "SPECTACULAR REALIZATION!" Years later, cosmologists would equate the significance of Guth's idea with the Big Bang theory itself.[7]

It would take two years, and a promotion from SLAC postdoc

to MIT professor, for Guth to work out his theory in full detail. In his epochal paper, "The Inflationary Universe: A Possible Solution to the Horizon and Flatness Problems," Guth sensed that he had planted the seeds for an entire field of cosmology.[8] But would these seeds blossom?

Indeed they would—beyond his wildest imagination.

FIELD OF DREAMS

The early 1980s milieu in which Guth's seeds germinated was a heady one. It was a remarkable epoch in physics, during which theories of condensed matter—the theory of gases, liquids, and solids—merged with theories of high energy physics—theories describing subatomic particles and the interactions between them.

Guth's spectacular realization was a proposed marriage between two seemingly unrelated branches of physics: condensed matter physics and particle physics. He asserted that there was a strange

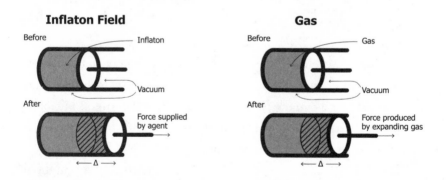

FIGURE 33. Guth's thought experiment illustrating the odd properties of the false vacuum. As the piston is pulled out of the cylinder, the volume increases. Since the energy density of the false vacuum remains constant and the energy increases, the extra energy is supplied by the force which pulls the piston out. Pulling on the negative pressure of the false vacuum requires energy, unlike the situation when an ordinary gas is in the cylinder (*right*). (ADAPTED FROM GUTH, "INFLATION," *PROCEEDINGS OF THE NATIONAL ACADEMIES OF SCIENCE* 90, NO. 11. © SHAFFER GRUBB)

entity called a *field* present in the very early universe. If this field had just the right properties, the universe could expand from microscopic scale to megascopic scale in the briefest flash of time, flatten out, and then seed tiny fluctuations so that there would one day be an Alan Guth to propose his inflationary universe. But we are getting ahead of ourselves. To understand inflation, we must first define what a field is.

A field is a mathematical tool describing how a physical quantity— say, temperature—changes from place to place in space. For example, the temperature of air in a room is a field: a set of numbers that changes with position, slightly lower near the window and higher near a heater. Fields like this, mere collections of numbers marking points in space, are known as scalar fields.

Guth showed mathematically that if the universe was filled with a special scalar field with just the right properties, then something almost magical could happen. He proposed that, in its infancy, immediately after the beginning of time, a scalar field he christened the inflaton existed. How the inflaton got there, Guth could not explain. He said that, in fact, it would turn out not to matter.

If the inflaton field was present from the beginning, as the universe expanded and cooled, it would have experienced some remarkable changes as well. Consider a cylinder filled with a gas that can be compressed or rarefied with a piston, as in fig. 33. If you were to suddenly increase the size of the cylinder, the density and temperature of the gas would correspondingly decrease. Equally, if you pulled the piston out fast enough, you would lower the temperature so much that the gas would liquefy: it would condense into a liquid, undergoing a process known to physicists as a phase transition.

Guth replaced the cylinder with the universe. The gas in the cylinder became the inflaton, and the role played by your hand pulling out the piston was replaced by what Guth called an unspecified "agent." The expansion, though it came about for unknown reasons, caused the inflaton to undergo a phase transition. But, as real life is not perfect, the condensation phase transition might not have occurred uniformly in all locations in the universe. For example, in our compressed air cylinder example, if the expansion were done extremely rapidly, there could be regions of the gas phase—in the form of bubbles—embedded within the liquid phase.

A STORM IN A CYLINDER

Aristotle said, "Nature abhors a vacuum." Like most things Aristotle said about physics, this was incorrect. A vacuum, to a physicist, is not a carpet-cleaner; technically speaking, a vacuum means the absence of matter and energy. Unlike Nature, Guth came to love, not abhor vacuums. His inflaton field required a strange kind of vacuum, known as a false vacuum—termed false not because it doesn't exist, but because it isn't a *true* vacuum.

A true vacuum is the state in which a system has the lowest amount of energy it can possibly have. For a ball rolling down a

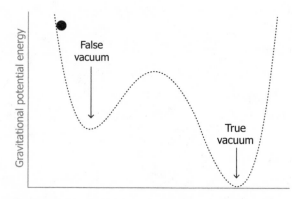

FIGURE 34. In a gravitational field, a ball rolling down a hill can come to a stable equilibrium position at rest either at the very bottom of the hill, where it has its lowest amount of gravitational potential energy, or in a divot at an intermediate value, which can have much higher energy. While the ball is still at rest in this intermediate value, known as a "false vacuum" state, it is unstable: if the ball is given more energy it will move away from the false vacuum state. (© SHAFFER GRUBB)

hill, the bottom of the hill is where the ball has the lowest amount of gravitational energy (fig. 34). That is its true vacuum state. But if the ball gets caught in a divot, though it might remain there at rest, it is not truly at the lowest energy state it could be in. It is in a false vacuum state, a place where its stored gravitational potential energy is at a "local" minimum.

To understand Guth's first version of inflation, let's consider a cylinder filled with two states of matter, liquid water with air bubbles inside. The water phase is stable but higher in energy and density than the air bubbles, which are in the true vacuum state. The bubbles move randomly amid the background fluid. Every once in a while, two bubbles collide, and when they do, they release energy stored in the bubble walls.

Guth proposed that the entire observable universe could have once been a single "patch" of the inflaton in its false vacuum phase—which, in the cylinder example, is the bubbly fluid filling it. As time evolved, bubbles of true vacuum formed. Being in the true vacuum state, the bubbles' energy density and pressure were highly uniform. This would have important ramifications later on, as the universe evolved, eventually causing the Big Bang's flatness and horizon problems to disappear.

Meanwhile, most of the pre-inflationary universe—the false vacuum state fluid filling the cylinder—expanded in size exponentially. Every once in a while true vacuum bubbles coalesced into one, and when they did, they released energy that then converted into matter (via Einstein's energy–mass equation, $E=mc^2$).

The Big Bang, in Guth's universe, boiled down to a sort of cosmic cauldron. It didn't matter how the liquid got into the cylinder, according to him, either: "The inflationary universe scenario begins with a patch of the universe somehow settling into a false vacuum state. The mechanism by which this happens has no effect on the later evolution."[9]

In Guth's model, our region of the universe began in a small patch of space, one perhaps smaller than an electron. As long as the original patch was in a false vacuum state, it expanded. So long as the expansion lasted at least a trillionth of a trillionth of a trillionth of a second, the universe inflated to macroscopic size.

Because the original patch was tiny, it was natural that it was uniform; there simply wasn't enough room for variation. Thus, the horizon and smoothness problems were solved. Our entire patch of the universe was part of the same initial false vacuum. When inflation then expanded that patch to become the observable universe, it also blew up the cosmos's radius of curvature. Any spatial curvature would be inflated enormously, like a balloon blown up to a huge size; though on the surface, you couldn't tell whether it was curved or not, as in fig. 31.

Thus was the Big Bang's flatness problem solved. *Without* inflation, the universe must have been "born" flat, with curvature differing from zero by, at most, one part in 10^{60}, a one followed by sixty zeros—extremely finely tuned indeed. *With* inflation, it could have had any value, including a radius of curvature near zero as Lemaître had suggested, and by the time we measured it, it would appear completely flat.

Last but not least, Guth had found the genesis of the Big Bang: the universe's expansion was fueled by the magical false vacuum. Once inflation ended, the universe's expansion rate slowed to the comparatively leisurely rate that Hubble observed. Not only had he explained the initial expansion of the universe, he also showed that "essentially all of the matter, energy, and entropy of the observed universe is produced by the expansion and subsequent decay of the false vacuum." For a while, Guth's effervescent vacuum seemed to satiate cosmologists everywhere. But soon, the bubble burst.

POP GOES THE COSMOS

The false vacuum fluid and the true vacuum bubbles were crucial to Guth's model. They each served a specific purpose: the false

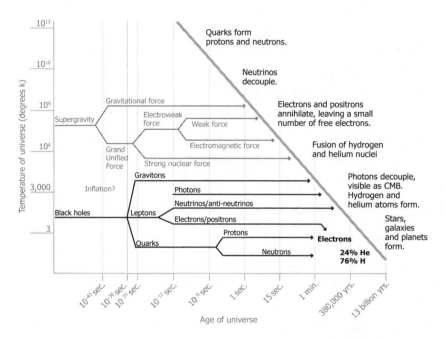

FIGURE 35. The thermal history of the universe. Inflation is thought to have begun when the universe was a trillionth of a trillionth of a trillionth of a second old. Other milestones in the history of the universe, the formation of the elements, and the formation of atoms, occurred much later (600 seconds later for nucleosynthesis and 380,000 years for chemosynthesis, the formation of the first atoms). (© SHAFFER GRUBB)

vacuum drove the exponential expansion and the true vacuum bubbles stored energy in their surfaces. When that surface tension energy was released it was converted to particles and radiation, such as leptons, quarks, photons, and neutrinos. All this happened before the universe was a trillionth of a trillionth of a trillionth of a second old. After that ripe old age, the temperature of the universe cooled as it expanded and eventually, once it was a few minutes old, nuclei could form. The rest, as they say, is cosmic history (fig. 35).

But inflation had a flaw, and it seemed fatal. The false vacuum growth rate was so prodigious that the universe expanded faster than the speed of light—and, therefore, the bubbles could never merge. Since the energy of a bubble is only released when it collides with another bubble—a process called percolation—the energy release that *would have* created matter in the universe could not occur. The bubbles couldn't find one another, leaving the universe sterile, cold, and barren—in clear violation of observations. And furthermore, under these conditions, inflation wouldn't stop—a conundrum soon called the "graceful exit problem." If the inflationary expansion never ended, it would be the *coup de grace* for the inflationary universe.

By now, you've probably noticed a trend: when cosmologists don't fully understand a theory, they say there's a problem, and then they come up with a new idea to solve the problem. But that solution brings up other problems. It seems like killing cockroaches—you can kill one, but there's another one right around the corner. So, now you're thinking, "Let me guess: someone else came up with another purely theoretical explanation solving the graceful exit problem. But then that caused another problem?" Well, yes, but that's what's so much fun about being an experimentalist: we get to be exterminators, providing pesticide in the form of data!

Just when things looked dim for Guth's model, in stepped Paul Steinhardt. Steinhardt was a high energy physicist who also understood the methods of condensed matter physics, the branch of physics that traffics in phase transitions. In retrospect, he was the ideal person to shepherd inflation through its darkest hour.

By chance, in 1980, Steinhardt heard a talk by Alan Guth just as the graceful exit problem seemed to be sealing inflation's doom. Steinhardt was fascinated. The mathematics behind inflation was intriguing, the interactions between particle physics and condensed matter physics were exciting, and the inability to gracefully terminate inflation was enervating. "It was the most exciting and most depressing talk I ever went to," Steinhardt recalled.[10] Surely he could find a solution, and fast. "I thought it would be a few weeks' digression," Steinhardt said to me in 2017, "but it's a digression that continues to this day!"

Steinhardt and his graduate student Andreas Albrecht got to work on the graceful exit problem. They realized they would need to kill Guth's darling, the abrupt bubble percolation process. They replaced it with a gentler process called the "slow roll." Instead of rapid bubble coalescence, their retooled model had the inflaton laconically converting into ordinary matter and radiation. The process still took only a trillionth of a trillionth of a trillionth of a second. Soon, a Russian physicist named Andrei Linde found a similar solution, lending credence to Steinhardt and Albrecht's proposal. The modified theory was called "new inflation."

Right after revamping inflation, Steinhardt, working with physicist Michael Turner at the University of Chicago, began to explore what a quantum theory of inflation might look like. Whereas "classical" fields such as temperature are familiar, in modern physics fields are *quantized*. Instead of smooth continuous functions, as in the case of temperature values, quantum fields can only take on certain discrete, or quantized, values.

Something astonishing occurred almost immediately. By quantizing the inflaton, Steinhardt and Turner made it roll in a discretized, jumpy (quantum) fashion instead of smoothly and continuously, as an ordinary ball would do. They received an unexpected dividend: new inflation unavoidably created tiny variations in space-time's curvature, tiny wrinkles.

The wrinkles came about because of the variation in the amount of inflation that took place. The amount varied from region to region. Wherever the inflaton field's energy was larger, even by random chance, the amount of expansion would be greater than in other regions. These regions of the universe would become voids, places where less material congregated and gravity was weaker on average. They would appear as hot spots in CMB temperature anisotropy maps. The opposite effect would happen where the inflaton energy density was lower; these would appear

FIGURE 36.

The temperature of the CMB at large angular scales is the result of the behavior of the primordial plasma, 380,000 years after the Big Bang. In the new inflationary model, the seeds for the large-scale temperature fluctuations observed by the COBE DMR experiment (*top*) arose due to the differing amounts of inflation experienced by different regions of the universe (*bottom*), a mere 10^{-36} seconds after the Big Bang. (© Shaffer Grubb)

as cold spots in CMB maps, as schematically depicted in fig. 36.[11] But, while new inflation provided a mechanism for fluctuations to arise, it couldn't explain how big, or rather how small, the fluctuations would be.

Still, new inflation was a great success. Finally, a theory of the beginning of the universe fixed more problems with the Big Bang model than it caused. Though the predicted pattern of CMB distortions went undiscovered until COBE found them in 1992, the pattern eventually would go down as inflation's first successful prediction—as opposed to a retrodiction, like the horizon and flatness problems that inflation was designed to solve.

Were there any unique predictions of new inflation, any do-or-die tests, that would elevate it beyond an edifice of elegant math to a complete theory of cosmogenesis? Cosmologists needed a "smoking gun" to prove inflation beyond a reasonable doubt, or else it would remain another just so story, like the biblical creation narrative which began this chapter.

GRAVITATIONAL WAVES

We cosmologists are a greedy lot. Discovering the CMB merely prodded us to inquire whether it was perfect or whether it had any anisotropy. When its anisotropy was found, cosmologists wanted to know how those tiny wrinkles were formed. Eventually we were emboldened to ask, "What came *before* the CMB existed?" Were there any fossil tracers of the Big Bang still around today that could take us back to "time zero" to test inflation? These relics needed to be primordial, ubiquitous, and resilient enough to last until today.

With infinite lifetimes, photons—the CMB's fossils—fit many of these criteria. But they were produced 380,000 years after the Big Bang—not even close to the beginning. Another stable and plentiful candidate, helium, was made in the universe's first twenty minutes. Deuterium was produced earlier still, around two minutes after the Big Bang—better, but not there yet. Cosmologists wanted to see back to the very beginning: just a tiny fraction of a second into the history of the universe, not twenty minutes.

The sought-after fossil messenger would come straight from central casting: gravitational waves, the variations in the force of gravity that travel at the speed of light and propagate continuously. Gravity seems like the strongest force of nature—especially if you've ever moved a couch—but in reality, it is the weakest of the four forces of nature. Yet this weakness begets gravity's strength. Gravity, it turns out, is especially mighty when it comes to testing inflation.

Soon after new inflation was proposed, cosmologists realized that the inflaton, being a quantum field, couldn't roll smoothly. Instead, it would jitter as it descended, a kind of drunken walk superimposed on the inexorable slow roll. In 1984, physicists Larry Abbot and Mark Wise showed that the chaotic motion of the new inflation model would inevitably generate gravitational waves—fossils not from the first twenty minutes, but from the first trillionth of a trillionth of a trillionth of a second after the Big Bang.[12]

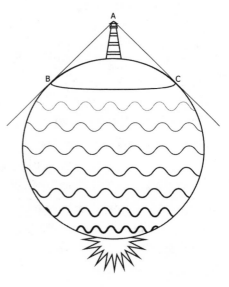

FIGURE 37.
An explosion on the opposite side of a watery world causes ripples that are detectable once the waves come into our horizon, the circle from B to C.
(© SHAFFER GRUBB)

Like photons, gravitational waves live forever. Unlike light, however, gravitational waves pervade everything—no type of matter in the universe can obstruct them. That's because gravitational waves are ripples in the fabric of space-time itself, whereas you can think of photons traveling *on* space-time. They would survive from the inflationary epoch to the time the CMB formed, at which time they'd distort the gravitational force field felt by photons.

The situation is similar to what you would see if you lived on a tiny, watery planet, as in fig. 37. On such a planet you would be able to detect an explosion occurring on the opposite side of the planet not by seeing it or hearing it, but by observing ripples on the surface of the ocean. The ripples play the role of gravitational waves. They carry information about the strength of the explosion (the amount of inflation, in the cosmological case), even though it is completely invisible. However, you need to wait until the ripples themselves come into view. For inflation, the gravitational waves would have rippled through the plasma that condensed to generate the CMB, imprinting it, and the CMB, with a telltale pattern revealing the strength of the inflationary "explosion" that took place long before the CMB itself was formed. If detected, that pattern would be powerful evidence for inflation; if it wasn't, it might be merely a wild goose chase.

LET THERE BE B-MODES

Just three years after Penzias and Wilson discovered the CMB, Cambridge astronomer Martin Rees predicted that the cosmic microwave background might be polarized if the ancient light possessed tiny deviations from perfect uniformity.[13] As we saw in chapter 6, as long as the light surrounding the electrons in the last-scattering surface plasma wasn't perfectly isotropic—perfectly uniform—the scattered light could become polarized, even if it was originally unpolarized; this is what DASI detected.

There are many ways to produce the requisite amount of anisotropy, but gravitational waves from inflation are certainly the most intriguing ones. Gravitational waves cause space-time to squish in one direction and spread in the perpendicular direction as they travel at the speed of light. Gravitational waves could have squished and spread the primordial plasma in such a way that photons would be slightly hotter along the axis of compression (producing blueshifted

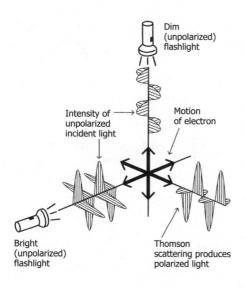

Dim (unpolarized) flashlight

Intensity of unpolarized incident light

Motion of electron

Bright (unpolarized) flashlight

Thomson scattering produces polarized light

FIGURE 38.
If the intensity of light seen by an electron in the primordial plasma is distributed non-uniformly, such that the two perpendicular axes have varying intensities, polarized light will emerge. This happens even if the incident light is unpolarized, as it is with a flashlight.

(© SHAFFER GRUBB)

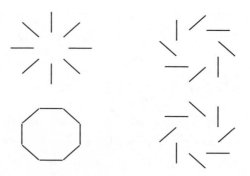

FIGURE 39. Pure E-mode (*left: top and bottom*) and pure B-mode (*right: top and bottom*) polarization patterns. Notice how B-modes have a swirly pattern while E-modes do not. E-modes can be caused by variations in the density of matter but, at large angular scales, B-modes can only be caused by gravitational waves. (© SHAFFER GRUBB)

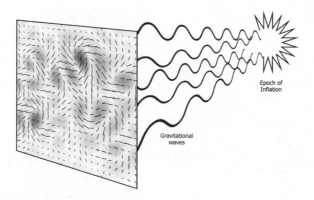

FIGURE 40. The squishing and squashing of space-time distorts the gravitational field of the last scattering surface. This distortion causes twisting patterns of CMB polarization. B-mode polarization can only be produced (at large angular scales) by gravitational waves, and it is thought that only inflation could have produced such primordial gravitational waves. (© SHAFFER GRUBB)

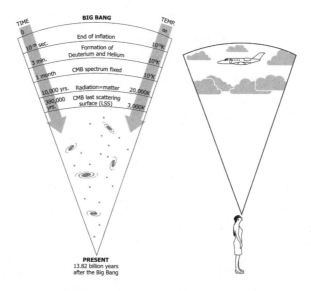

FIGURE 41. The CMB's last scattering surface and a cloudy marine layer are both opaque to light due to scattering (by electrons in the last scattering surface plasma, by water droplets in the marine layer). However, the existence of phenomena earlier in time (in the CMB's case) or farther above (in the marine layer's case) can be inferred by other means. A plane flying above the marine layer can be detected by the sound it makes. In the CMB's case, a signal from long before the CMB was produced can be examined using the polarization of the CMB caused by gravitational waves. (© SHAFFER GRUBB)

photons) and slightly cooler at a right angle to this axis (producing redshifted photons), as shown in fig. 38.

Alex Polnarev, a Russian physicist who later became a mentor to me, recognized in 1985 that the alternating asymmetric shear and stretching of space-time caused by gravitational waves would inevitably induce a twisting, swirling pattern in the orientation of the ancient light's polarization. Polnarev conjectured that, in principle, it could even be detectable. This unique polarization pattern, depicted in fig. 39, would be, in effect, inflation's signature; it would exist if and only if cosmological gravitational waves had interacted with the primordial plasma whose imprint the CMB carries. It was later called B-mode polarization.[14]

If there were enough of these primordial gravitational waves, B-mode polarization would appear in the CMB—if only we could see it (fig. 40). It would be incontrovertible evidence for inflation.

On paper, the story of inflation was complete. It was a true crime drama whose prime suspect, the inflaton, and its hideaway, the first fraction of a second after time began, had both been identified (fig. 41). There was even, finally, a smoking gun—gravitational waves—that would provide hard evidence of what had taken place. The smoke from the gun was the glorious swirl of B-modes.

This was what I so desperately wanted to capture with the telescope I designed as a postdoc. Soon, I hoped, BICEP would become reading glasses for the greatest story ever told.

PLATE 1. The Nobel medal of the Royal Swedish Academy of Sciences depicts Alfred Nobel on the obverse side. The reverse side depicts Mother Nature emerging from clouds, holding a cornucopia. The veil covering Mother Nature's face is removed by the Genius of Science. (© THE NOBEL FOUNDATION)

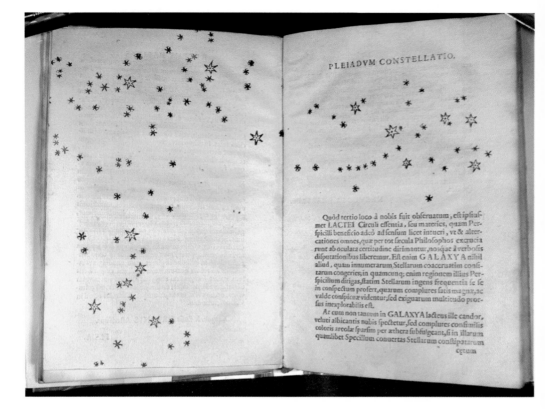

PLATE 2. Galileo Galilei's sketch and description of the Pleiades from *Sidereus Nuncius*. (FROM THE COLLECTION OF JAY PASACHOFF)

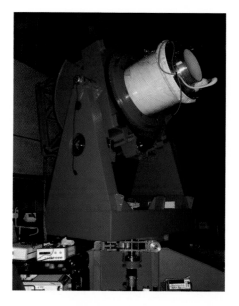

PLATE 3. *Left*: The BICEP telescope on its mount in 2005, before deployment to the South Pole. (BRIAN KEATING) *Bottom*: The BICEP telescope and its environs, inside the Dark Sector Laboratory at the South Pole. (JEFFREY DONENFELD)

PLATE 4. The BICEP telescope is located under the aluminum "cone" on the roof of the U.S. Amundsen Scott South Pole Station's Dark Sector Laboratory. The 10-meter-diameter South Pole Telescope is visible at left. (JEFFREY DONENFELD)

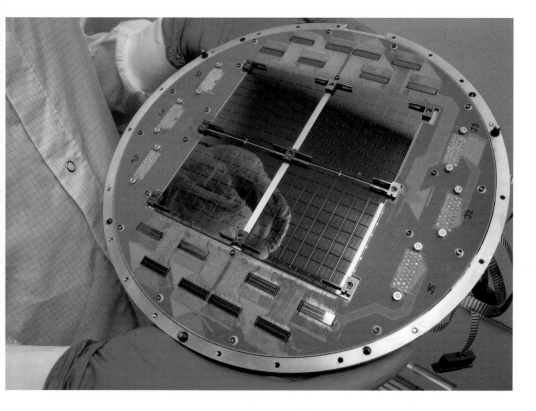

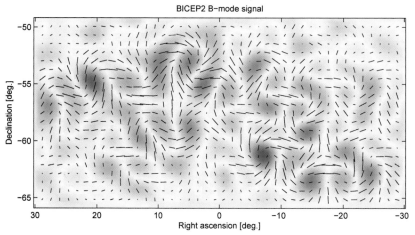

PLATE 5. *Top*: The BICEP2 focal plane array consists of 512 polarization-sensitive superconducting bolometers. (NASA) *Bottom*: BICEP2 map of B-mode polarization. (BICEP2 COLLABORATION)

PLATE 6. *Top* (left to right): Paolo de Bernardis, Paul Richards, and Andrew E. Lange at the 2009 Dan David Award festivities in Tel Aviv, Israel. (DAN DAVID FOUNDATION) *Bottom*: The Planck satellite and its measurement of the temperature anisotropy of the cosmic microwave background radiation. (EUROPEAN SPACE AGENCY)

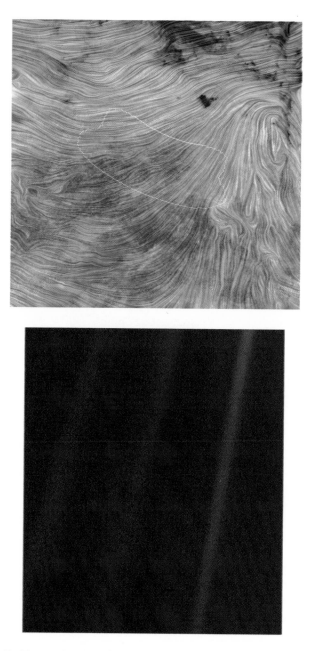

PLATE 7. *Top*: BICEP2's observing region as seen by ESA's Planck satellite at microwave and sub-millimeter wavelengths. The color scale represents the thermal emission from interstellar dust. The "streamlines" indicate the orientation of the Milky Way galaxy's magnetic field, measured using polarized light emitted by the dust. (EUROPEAN SPACE AGENCY) *Bottom*: The "Pale Blue Dot" photograph, showing Earth as described by Carl Sagan as "a mote of dust suspended on a sunbeam," taken on February 14, 1990, by Voyager 1. (NASA)

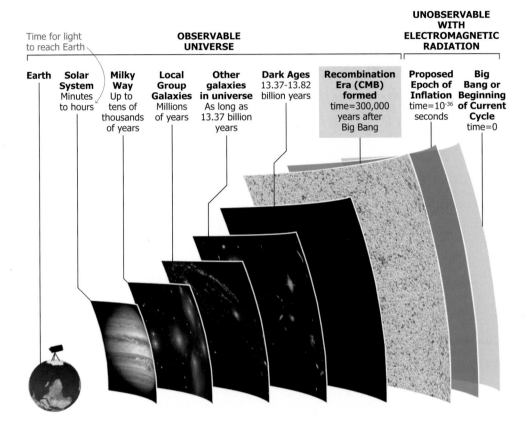

| Time for light to reach Earth | OBSERVABLE UNIVERSE | | | | | | UNOBSERVABLE WITH ELECTROMAGNETIC RADIATION | |

| Earth | Solar System Minutes to hours | Milky Way Up to tens of thousands of years | Local Group Galaxies Millions of years | Other galaxies in universe As long as 13.37 billion years | Dark Ages 13.37-13.82 billion years | Recombination Era (CMB) formed time=300,000 years after Big Bang | Proposed Epoch of Inflation time=10^{-36} seconds | Big Bang or Beginning of Current Cycle time=0 |

PLATE 8. The standard model of cosmology, from the beginning of time to the present day. Major epochs in the history of the universe are depicted, ranging from the possible inflationary epoch to the formation of galaxies, to the formation of stars and planets in the late universe. (© SHAFFER GRUBB)

Chapter 8

BICEP: THE ULTIMATE TIME MACHINE

"All life is an experiment. The more experiments you make the better."

RALPH WALDO EMERSON

COSMIC MICROWAVE BACKGROUND EXPERIMENTS AND inflation seemed to be in cahoots. Experimental findings would come in, always consistent with inflation. Penzias and Wilson's discovery suggested a universe that was evolving, cooling, and expanding; inflation provided the reason. COBE found tiny fluctuations in the otherwise featureless CMB, a discovery that inflation presaged. BOOMERanG unequivocally showed the universe was completely flat, and again inflationary cosmologists nodded knowingly. The CMB seemed like a Stockholm slot machine paying out in golden prizes.

The expansion, flatness, and homogeneity that were found in CMB data were all necessary, but not sufficient, conditions to prove inflation. That's why B-mode polarization was so exciting; it was my best chance for the Nobel Prize.

In fact, detecting B-mode polarization should yield not one but *three* Nobel Prizes, I reasoned. One would, of course, be awarded for spying B-modes, which would be definitive evidence for inflation and its new scalar field, the inflaton. (Discovery of a new scalar field resulted in the 2013 Nobel Prize for the Higgs boson.) The second one would be for the indirect detection of primordial gravitational waves. And the third, but certainly not least, would go to whoever proved that primordial B-modes represented not just gravitational waves, but gravitational waves that were produced from quantum fluctuations in the inflaton. It would be our first, and perhaps only, evidence for quantum gravity.[1] It wasn't just cosmology. It wasn't just experimental cosmology. It was experimental *quantum* cosmology, a discovery for the ages.

If only I could do it. But of course, it wouldn't be me alone; no one goes to Stockholm solo.

COLLABORATE AND LISTEN!

"I hear you like to play tennis." It was the fall of 2000, just a few months after I had been booted out of Stanford and settled into Pasadena. I had finally gotten up the nerve to ask Dr. Jamie Bock to get together outside of work. Tall, brilliant, and equipped with a racket-length resumé, he was intimidating. He looked every bit the part of the president of an East Coast fraternity, right down to the sporty Mazda Miata he drove to Caltech.

After attending Duke University, Bock went to UC Berkeley where, as a graduate student, he built a small rocket-borne telescope in Andrew Lange's lab. He then headed to NASA's Jet Propulsion Laboratory as a project scientist, caught in the no-man's-land between postdoc and professor. Unfortunately, the primary obstacle to his becoming a professor at Caltech was Lange. Most universities, even wealthy ones with enormous endowments like Caltech, cannot justify having two professors working on the same project. And so Bock was caught in the dreaded "soft money" position, having to propose for funding each year, seemingly permanently out of contention for a faculty position and left at the crossing rails of the tenure track.

But the job seemed to suit Bock. He loved working on technical details. While he wasn't the idea man Lange was, he could find the most advanced, elegant, technological solution to every problem Lange got him into. He was the Steve Wozniak to Lange's Steve Jobs, the consummate craftsman toiling out of sight.

By the time I met him, Bock was already a legend in our community. He had been the key player in many of Lange's successful experiments, most notably BOOMERanG, the balloon-borne instrument that discovered the flatness of the universe. BOOMERanG catapulted Lange to worldwide fame, as well as bringing Lange to Stanford at the very moment when I was on the way out. Although Bock had designed and built BOOMERanG's ingenious spiderweb bolometers, you might not know of it from a distance; as a professor, it was Lange who got most of the credit.

In 2000, BOOMERanG had confirmed what theoreticians suspected: the universe was free of curvature, suggesting either that the universe was created perfectly flat or that some mechanism like inflation had made it so. BOOMERanG, combined with COBE's discovery of the smoothness of the universe, meant that most of the low-hanging CMB inflationary fruit had been picked. But still missing was *data* explaining why the universe was so flat, fecund, and fit for life. BOOMERanG wasn't designed to detect B-mode polarization.

To build a CMB experiment at Caltech required Andrew Lange's buy-in. While Lange had been impressed by my bold pitch for a new telescope design during my job interview, it was only enough to get me a job in his group. At first, my post was more probationary than permanent; after all, he'd only heard about me *after* I was fired by his former postdoc Sarah Church. I still needed to earn the benediction of experimental cosmology's high priest. There was a lot of competition for his attention. He had six other postdocs, all of whom felt fortunate to be in his lab. Every one of us clamored for his blessing someday: a glowing letter of recommendation, a guaranteed ticket to a faculty job. There had to be a way for me to stand out from the horde of postdocs; I needed an in. I hoped it would be Bock, Andrew's essential sidekick.

I had played tennis in high school, though my skills had deteriorated in the dozen intervening years. Soon after landing at Caltech, I saw that Bock kept a racket by his desk. Across from Lange's laboratory on California Boulevard in Pasadena were a few tennis courts that were often free on Thursday nights. In fact, it wasn't a great feat to get exclusive use of *any* of Caltech's athletic facilities on most days.

Bock and I agreed to play weekly games. It was a rare chance, at an ultracompetitive place like Caltech, to break a sweat physically. Unlike at many athletic bouts at Caltech, the home team (Bock) almost always won. But I kept it interesting, and he never canceled our matches. Between sets, I'd try out my ideas for new experiments.

Then, after a few weeks, I won a match. Emboldened, I shamelessly appealed to his passion for tennis. "Jamie," I said, "what if I told you we could measure inflationary gravitational waves using the CMB's B-mode polarization? Wouldn't that be the biggest of all grand slams?"

At first, he seemed unwilling to add insult to my tennis-bruised ego, and barely humored my idea. A taciturn "Uh huh, okay, sure," was all he could muster. But I persisted. The idea of testing inflation was intriguing, but unless there was some technical meat for him to bite into, my one chance for Nobel glory would pass me by.

I explained that there was a convergence, using the new technology he and Lange had been working on. The two cosmologists had been tasked with building bolometers for the upcoming Planck satellite that, unlike their previous spiderweb detectors, would be sensitive to the CMB's polarization. Those new sensors, combined with the simple telescope design my PhD thesis experiment (POLAR) had used, could measure the inflationary B-modes. Best of all, there'd be a lot of nitty-gritty technical work to do.

I had made my approach, but I would need help convincing him. Fortunately, my arrival at Caltech coincided with that of Professor Marc Kamionkowski, the renowned young theoretical physicist who coauthored the "Polarization Pursuers' Guide" that had so delightfully distracted me from my postdoctoral duties in Sarah Church's Stanford lab the year before. Kamionkowski helped me make some plots showing how we could use a small telescope to hunt for B-modes.

This approach was appealing for a couple of reasons. A small refracting telescope had worked for Galileo; his telescope was just big enough to see the moons of Jupiter, so crucial to his anti-geocentric arguments. Every telescope has a finite resolution: the smallest astronomical object it can detect. Resolution is determined by the diameter of the telescope's lens or mirror, and the wavelength of light it collects. Telescope diameter is to astronomers what acreage is to realtors—a prized commodity, second only to location. But astronomical real estate comes at a cost, a cost which rises not in proportion to the telescope's diameter but to its collecting area (or, diameter squared). A telescope that is 60 cm (2 feet) in diameter costs four times as much as a telescope that is 30 cm (1 foot) in diameter. Smart astronomers build telescopes just big enough to resolve the signals they seek.

With Kamionkowski's aid, I convinced Bock. Now, I just needed Bock to help me propose the idea to Lange. We worked on an elevator-style pitch: we would capture the Big Bang's baby picture using a camera that Lange's hero and mine, Galileo, would recognize: a refractor. The day came to pitch it to Lange. As we finished our brief spiel, Lange erupted, "What you're proposing will cost millions of dollars!"

My heart stopped. Lange was a pragmatist, not known for doing things just because they were cool, especially when they were so expensive. There had to be a justification for such a princely sum. I was crushed; my dream had ended before it began. But then Lange took a breath, and added, "But that's okay . . . that's okay!" He had bought it.

His blessing was all that mattered. We had our marching orders and went to work. After making sure the telescope could really do the job, I set out to devise an evocative name for the experiment. I came up with "Background Imaging of Cosmic Extragalactic Polarization" or BICEP—the biceps being the muscle behind the exercises called curls, which was a nod to the alternate name coined by Kamionkowski, Arthur Kosowsky, and Albert Stebbins for the pinwheel-like polarization pattern more commonly known as B-modes.

Our second priority was money. Whereas Galileo had fretted that his telescope design would be scooped if he publicized it, we had the opposite problem: we desperately wanted attention. Lange, Bock, and I wrote our first funding proposal in 2002. Caltech president (and Nobel laureate in physiology or medicine) David Baltimore became our Cosimo de' Medici when he underwrote it using his president's discretionary fund. I'm sure that without this special money, the experiment would not have happened; astronomers always thought bigger telescopes were better. BICEP challenged this paradigm and was likely too speculative for the more staid federal agencies to back without some shared risk. Who could blame them? Even Lange often joked with us that pursuing B-modes might turn out to be the ultimate wild goose chase.

Soon after we got funded, Andrew Lange sent me on a "speaking tour" to draw attention to BICEP. He even forced me to go on an all-expenses-paid trip to speak at a conference in Kona, Hawaii. This voyage led to the first academic paper describing a new refracting telescope since the Yerkes Observatory had opened over a century earlier.[2]

BICEP: SOME ASSEMBLY REQUIRED

BICEP took five years and two million dollars to build. You can, however, build a polarimeter simply by donning polarized sunglasses, looking at the zenith at sunset, and spinning around in place. Because light is polarized when it scatters off air molecules, you'll notice the brightness of the sky varies *twice*, from bright to dark to bright to dark, every time you spin a full circle (fig. 42). This twofold brightness variation is the signature of polarization.

Like your sunglasses polarimeter, all polarimeters have four fea-

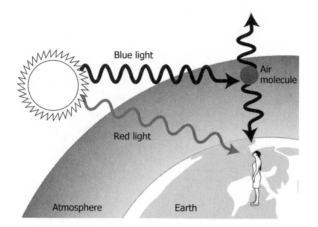

FIGURE 42. Polarization of sunlight. Only the scattered light is polarized, as you can verify by using a pair of polarizing sunglasses and spinning around in place looking at the zenith at sunset; the light's intensity varies twice for each single time you spin around. This effect does not occur for the directly transmitted light from the Sun because it's unpolarized. (© SHAFFER GRUBB)

tures in common: optics (for you, the lenses of your eyes), a polarizing filter (the sunglasses) to separate vertically polarized light from horizontally polarized light, detectors (your retinas), and a polarization modulator (your legs keeping you spinning) that causes the intensity of the light through each of the polarizing filters to vary predictably.

So too did BICEP feature these same four essential polarimeter elements. BICEP's optics were 30 cm (1-foot) diameter lenses made of high-density polyethylene, the same material used in milk jugs. Though these containers appear opaque to the eye, they transmit microwaves almost perfectly. The two lenses produced clear vision over a huge field of view nearly twenty degrees wide—equivalent to two fists held at arm's length.

BICEP's polarizing filters and detector were combined into what were called polarization sensitive bolometers, or PSBs (fig. 43). We didn't require separate sunglasses and retinas; the detectors themselves were intrinsically sensitive to the polarization state of light.

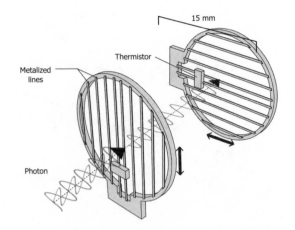

FIGURE 43. Schematic diagram of BICEP's polarization-sensitive bolometers. BICEP's bolometers combined a polarization filter (metal lines deposited on a thin silicon nitride film) and two detectors, one per polarization state, to detect CMB polarization. (© SHAFFER GRUBB)

Each of BICEP's pixels had two bolometers: one for horizontal polarization and another for vertical polarization.

In truth, each of your retinas is not a single "detector." A retina has about 6 million cone cells, which register the frequency of light and can distinguish between millions of colors. And the retina's 100 million rod cells allow humans to perceive the faintest light source, even a single photon.[3] BICEP was looking for tiny variations in heat (the CMB), not visible light; therefore, our "microwave retinas" needed to be cold—as cold as possible, for as long as possible. And just as a human retina has numerous rod and cone detectors, we needed multiple detectors for BICEP. Compared to the retina or even a smartphone camera, BICEP's detector count—98—seems pitiful. But if we were lucky, our pixels would capture waves of gravity coursing through the oldest light there is. No phone, no matter how smart, could even come close to taking that picture.

To detect the faint CMB heat at all, the detectors needed to be cooled to just a quarter of a degree Celsius above absolute zero. Here,

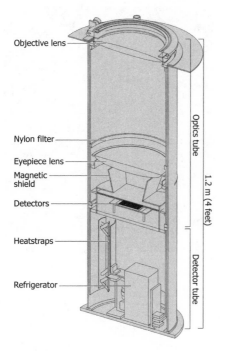

Objective lens

Nylon filter

Eyepiece lens

Magnetic shield

Detectors

Heatstraps

Refrigerator

Optics tube

Detector tube

1.2 m (4 feet)

FIGURE 44.
Cross-section of BICEP's cryogenic telescope, showing the location of its principal components. The entire telescope was held under vacuum and cooled by liquid helium to −269°C (−453°F).
(© SHAFFER GRUBB)

once again, BICEP's Lilliputian size was its biggest asset. Since it was small, barely 1.5 m (5 feet) long, the entire BICEP telescope—optics, polarizing filters/detectors—could be put inside what was essentially a giant thermos. BICEP's thermos was a cylindrical vessel just large enough to contain all the optical elements and hold them at a pressure less than a millionth of what you feel at sea level (fig. 44).

Keeping the pressure inside low was crucial; if there were too many air molecules inside the cryostat they'd quickly rob heat from the walls of the thermos, bringing unwanted heat into the detectors and rendering them useless. The thermos had two chambers within it filled with liquid helium. A dedicated refrigerator held a liquid form of an isotope of helium, called liquid helium-3. Ordinary liquid helium got BICEP to about 3 kelvin, and helium-3 helped it reach 0.25 kelvin. For the first time in human history, we had cooled an entire telescope to the temperature of interstellar space. Our cool telescope would have impressed Galileo, who knew a thing or two about thermometry, as well, of course, as being an expert in refracting telescopes.

The final hallmark of all polarimeters is the polarization modulator. For BICEP, that duty was performed by a 10,000-pound mount capable of controlling the complex rotations of the telescope with fantastic precision. For more than a year we assembled and reassembled BICEP, putting it through its paces (fig. 45). It was far easier to make sure it was working in Pasadena than wherever it would end up.

BICEP was an experiment, and no experiment works immediately. Once the pieces of a new experiment are put together, we experimentalists spend most of our time puzzling over the instrument's imperfections in hopes of reducing "errors"—although that term does a poor job of capturing the nature of the work.

These errors aren't mistakes. Instead, they are the inevitable byproduct of doing experiments in an imperfect, noisy world. An error is simply the difference between what you measure and the value it would be if the instrument was perfect (a value that you don't know). The *uncertainty* of a measurement is the range that likely contains the correct value of the quantity. After the experiment has run and we have measurements, experimentalists ascribe a *confidence level* to those measurements, standing for the likelihood that the true, would-be perfect value of the measurement lies within the uncertainty bounds we set.

Two types of errors deserve special attention. First, an instrument's *systematic error* causes biased measurements. For example, a thermometer that always reads ten degrees too high needs to be corrected—a process called calibration—by, say, using another, previously calibrated thermometer and comparing the two. Afterward, this offset can be simply subtracted. The second special type of error is just called noise; it is due to the random variation of all measured quantities. Fortunately, noise behaves predictably and can be reduced by repeating the measurement many times to average it out.

The better the experiment, the more immune it is to systematic errors and noise. There is no shortage of contaminants; they can arise from the instrument itself, its surroundings, or the universe. Systematic errors reside in the measurement itself, in the way you built the instrument, in the way you did the data analysis, or something physical other than the quantity you'd like to measure. Experimentalists build the lowest-noise instrument possible, and then spend most of their time looking to remove systematics. Their aim is to maximize both the precision (the total of all the random noise errors present) and the accuracy (how close a measured value is to the true value) of their measurements. (Note that accuracy doesn't imply precision, and vice versa.)

When scientists speak about the quality of their measurements, they use a quantity called the signal-to-noise ratio, denoted SNR.[4] The SNR encapsulates how significant the measurement is— meaning, how confident we are that it is real, not just a random fluctuation. A low SNR indicates low confidence in the measured signal, suggesting it may just be a statistical fluke. A high SNR suggests high confidence in the reality of the signal. Typically, to claim credit for discovery of a new cosmic phenomenon, the signal must be at least three times bigger than the noise. This was the case with Penzias and Wilson: they measured 3.5 kelvin for the new celestial signal—the cosmic microwave background—with an uncertainty of about 1 kelvin. That meant their SNR was 3.5. Confirmation came quickly as other experimentalists saw comparable signals.

But, recall, Penzias and Wilson weren't looking to prove anything—they were looking for nothing. We, on BICEP, had a specific target in mind, even though we didn't know whether we could hit it. Penzias and Wilson's signal level turned out to be larger than their noise level. Our signal level was completely unknown. For all we knew, it could be zero, which would always result in SNR = 0.

RADAR LOVE

Penzias and Wilson had one more advantage over BICEP: they were using the same 6 m (20-foot) diameter microwave telescope antenna that Ed Ohm had used. And they had a killer app, technology which their unknowing competitor Robert Dicke had invented decades earlier when he was developing radar as part of the effort to win World War II. Dicke devised a modulator, now called a Dicke switch, which finally removed the systematic errors that had plagued Ohm's use of the telescope. The lesson was clear: to get rid of systematics, we'd need to build our own "switch." We weren't about to make the same mistake Ohm had.

Lange and I spent hours in the lab figuring out how BICEP's thermos was behaving and how its PSB detectors were working. He would delight in its peccadillos, marveling at the unexpected complexity of the instrument. I felt a bit guilty having sold him on it by pitching its simplicity. "BICEP's behavior is more complicated than all of cosmology," he joked at one point, admonishing me not to get depressed over odd phenomena the telescope sometimes showed. Then he added one of his catchphrases: "A new experiment of this kind is like falling in love. There's no way anybody can stop it."[5]

He and I designed new devices that worked like a polarized Dicke switch. We called them Faraday Rotation Modulators. And voilà, they worked! So well, in fact, that we sought a patent: the first patent for each of us.[6] We only had 354 more to go to catch up with Alfred Nobel.

We still needed to know where to look in order to capture the polarization imprimatur of inflationary gravitational waves. We knew that the B-mode signal, if it existed, would be brightest at two-degree angular scales, meaning the pattern would subtend an area on the sky about the size of your thumb held at arm's length. But we couldn't just cover a single two-degree wide patch of sky. Say we saw something in a single two-degree patch, how could we be sure it wasn't a fluke fluctuation? We would need to measure many two-degree patches.

With the help of fellow Caltech postdoc Eric Hivon, a theoretician, we calculated that BICEP needed to cover an area of about 1000 square degrees on the sky. A patch of sky that size could fit hundreds of B-modes, allowing an inarguably significant detection.

Where could we find such a patch of sky? We'd need a region with little or no contamination from the Milky Way galaxy. The Milky Way, as we now know, contains gas and dust—that cosmic dirt that has messed with astronomers since Galileo. In 2001, no good surveys had yet been done of the polarization of the sky at these scales, at these frequencies. We'd need to locate a suitable region ourselves, which might be tricky. Unlike Galileo, we couldn't just point our telescope at some bright object in the heavens, like Jupiter. Worse yet, our signal was microwave light; we couldn't even *see* a good target to look at. Where we cast our cosmic gaze turned out to have consequences that echo even to this day.

LOCATION, LOCATION, LOCATION

We knew the inflationary gravitational-wave background signal would be small. It would take years to find it, if it even existed. We also knew the sky holds many sources of microwaves, which could masquerade as inflationary B-modes, such as from the atmosphere and the landscape surrounding the BICEP telescope. Even the Milky

Way galaxy could mess with our measurements; it was abuzz with microwaves, some coming from glowing dust irradiated by starlight, others from electrons whirling around the galaxy's magnetic fields.

My group began to look for some prime cosmic real estate. We found what seemed to be the Goldilocks patch—not too big, not too small, and just right to stare at for years.

Next, we needed the perfect Earth-bound observation post from which to aim BICEP. We knew we couldn't do the experiment from California, since the amount of water vapor in the atmosphere, even over the deserts of Southern California, is far too high. Water, while crucial to life, makes a cosmologist's life hard—it is a very efficient microwave absorber. In an ideal world, we'd get off the world's surface altogether, above *all* of the Earth's atmosphere. But to do that, we would need a satellite and close to a billion dollars.

For less flush scientists, unable to afford a space-based telescope, the South Pole comes a very close second. The South Pole is technically a high-altitude desert (at upward of 2.7 km/9,000 feet above sea level), with tiny amounts of precipitation each year. The snow you see in pictures of the Pole is mostly imported, blown in from the coastal regions of the continent over eons; that's how cold it is. That cold helped our telescope in another way: it reduced some of the excess heat that could contaminate our measurements. The sky is dark six months of the year. Best of all, the South Pole was the site of the DASI telescope, which had detected the CMB's polarization for the first time, in 2002. It was an ideal choice.

With help from one of Lange's former grad students, a UC Berkeley cosmologist named Bill Holzapfel, Bock, Lange, and I got funding from the NSF's Office of Polar Programs (OPP) to take BICEP to the U.S. Amundsen–Scott South Pole Station. Getting funding from the government was a huge boost for all of us—especially me, because it was my first time leading such an endeavor. It felt awesome.

Things got even better when OPP generously gave us most of a building, ominously named the Dark Sector Laboratory, located about half a mile from the South Pole itself. (I was relieved to find out it was called the Dark Sector not because of any affiliation with Darth Vader, but because it was a region where no radio communication of any kind was allowed, meaning it was "radio-dark.") With everything in place, by late 2005, BICEP was ready to leave the nest and fly south for many winters.

Slowly it dawned on me: to get to Stockholm, I first had to go to the South Pole. It wasn't exactly a trip I was dying to make. In fact, I was more worried about dying in the attempt, as the English explorer Robert Scott and his team had nearly a century before. But I had come too far to turn back now; I had become a prisoner of my dreams. Emulating the heroic Brits, I set my upper lip to "stiff," and set off on the adventure of a lifetime, hoping it wouldn't also be my last.

Chapter 9

HEROES OF FIRE,
HEROES OF ICE

*"I may as well confess that I had no
predilection for polar exploration."*

ROBERT FALCON SCOTT, 1904

ANTARCTICA. THE VERY WORD STILL SENDS
shivers up my spine. It's the coldest, windiest, least populated, and most remote place on Earth. There are scientific outposts in Antarctica inhabited by fewer people than currently abide on the International Space Station. Travel in and out of the continent is extremely restricted. Access to the South Pole is rarer still.

As I prepared for the journey, I learned as much as I could about the heroic explorers of the twentieth century. The South Pole, the bull's-eye at the bottom of the planet, was the target of many explorers after Antarctica itself was discovered. They pursued it passionately, despite its dangers. I felt so unworthy compared to them. In my research, I noticed comparisons between the heroes of exploration and my scientific heroes, those who pursue the Nobel Prize with a passion equal to that of polar explorers.

Winning, in science or exploration, confers international glory and everlasting fame. Both endeavors lie at the extreme of human intellectual and physical capability. As with the first polar explorers, a few early laureates paid for the prize with much physical suffering—most notably Pierre and Marie Curie.[1] And both endeavors are fraught with isolation, competition, and even treachery.

In 1901, the year Wilhelm Röntgen received the first Nobel Prize in Physics, Robert Falcon Scott made his first voyage to Antarctica, hoping to be the first man to reach the South Pole. Though he did not reach the Pole—he got within five hundred miles before being forced to turn back due to scurvy—his team learned much about the geography and climate of the forbidden continent. He left Antarctica in 1904 chastened, but resolved to return again.

In the months leading up to my deployment to "the Ice" in December 2005, I was consumed by the diary Scott wrote during his second, ill-fated trip to the Pole.[2] He reminded me of a modern-day scientific leader, a true "principal investigator." Like us, he meticulously stockpiled supply caches and assembled a crew (a motley band of brigands and brainiacs). He also shared our abiding faith in science, even in recently invented disciplines such as polar meteorology.[3]

In contrast, Scott's competitor, the Norwegian Roald Amundsen, led his men on a furious sled-dog-powered dash to the Pole, unencumbered by the obligations and logistical demands of scientific exploration. The poor dogs would later become an après-ski meal for the Norwegians. As Scott prepared to set out for the Pole, he found, to his shock, that Amundsen, who had missed out on being the first to reach the North Pole, was racing him to the South Pole. Fear kicked in. Scott made what many believe was his fatal mistake. Instead of turning back as he'd done in 1904, he pressed on, desperate to beat the Norwegians. This meant that Scott's team would have to go faster, consuming their food and fuel more quickly, perhaps even before they could reach their life-saving supplies. In retrospect, it wasn't the cold that did them in. It was fear: fear of losing the polar prize, the trophy that had become inextricably linked to Scott's identity.

RACE TO THE BOTTOM

Luckily for me, the National Science Foundation's expert Office of Polar Programs was in charge of planning my trip. I would basically be chauffeured to the South Pole, on military aircraft, the greatest hardship being the shock I had when learning that passengers are called "self-loading cargo" on the plane's manifest.

My memories of my first polar voyage in 2005 are still vivid, thanks to the meticulous journal I kept. Like Robert Falcon Scott, I didn't know when, or if, I'd be back. But unlike Scott's, my journal entries tell a tale lacking in all but the slightest bit of privation. I set out on December 10, 2005, the 109th anniversary of Alfred Nobel's death, with two suitcases: one held kosher food, my food cache for a month-long "expedition." The other was for now mostly empty but would soon be stuffed to the gills with clothes for Antarctica. Like all mothers, mine gave me sage advice: "Bring a jacket!" I foolishly dismissed her counsel, telling her, "Mom, they will give me a jacket. In fact, they'll give me two if I want: one thick canvas Carhartt and one goose-down jacket." The latter was bright red with a reflector on the back to aid search-and-rescue parties in finding my frozen, lifeless body on the ice, if need be. Being a dutiful son, I neglected to tell her that last part.

After a short hop from San Diego to Los Angeles, I was off to the departure point for Antarctic explorers throughout history: Christchurch, New Zealand, a lovely little city modeled after Stratford-on-Avon, England. The city's river is even called the Avon, and one can pass the day punting on it in a small wooden gondola. Robert Scott deployed from Christchurch on his final voyage to the South Pole, and his wife carved a huge marble sculpture of him five years after he failed to return. Poignantly, it shows Scott walking north, as if he were trudging back to her in London.

A night in Christchurch allows just enough time to check in for your flight and collect your Antarctic gear, for you will depart imminently and there is no Walmart where you're going. Not long into the whole affair you're inundated by acronyms. "Go to the CDC to get your ECW. Then, and only then, are you cleared from CHC to MCM." (That barrage stood for Clothing Distribution Center, Extreme Cold Weather gear, Christchurch, and McMurdo—the last being the U.S. base on the coast of Antarctica, directly south of New Zealand.) I was anxious because I learned I might be flying to Antarctica on a C-130, an old military cargo plane also known as a Hercules, instead of the jet-powered C-17 Globemaster, which takes half as long to jump the frozen puddle.

Departure day dawned early (at least for an astronomer) with a 5:30 a.m. wakeup call from the polite hostel owner, who was also its entire staff. I unpacked the mountain of clothes I had packed only the day before and got dressed in ECW survival gear, mandatory for the flight down to Antarctica. Garbed in forty pounds of clothing— mind you, in summer—I went to get a final cup of real coffee at the Antarctic Centre at the Christchurch airport. The Centre is a wonderful little museum designed to emulate a fraction of the thrills of Antarctica. I decided not to take the simulation tour, skipping a ride on a Snowcat—a motorized sledge—in simulated Antarctic conditions of −18°C (0°F). I figured I'd get plenty of real adventure on the Ice. After the cheerful lads of the Kiwi air force screened us for illegal drugs—seemingly the only forbidden items on the entire continent—I was ready to go.

The C-130, a four-engine behemoth first deployed in the 1950s, is the turboprop workhorse of the continent. Inside, spartan accommodations await, as befits a plane made to drop a platoon under silk canopies. A beast of burden that has served in a dozen wars since being introduced six decades ago, it has no reclinable seats. In fact, it has no true seats whatsoever, just a hard bench or a cargo net to sit in for nine hours. Don't bother looking for a lavatory; there aren't any, just a couple of five-gallon buckets, affectionately known as the honeypots, hiding out in the back of the aircraft.

There are only two passenger windows in the cabin, and of course no galley, never mind any in-flight meal service. The flight attendant is a surly sergeant known as the cargomaster, whose only interest in scientists (whom he derisively calls "Beakers") is that they keep their damn hands away from the jump button—the indicator notifying the pilots to open the ramp so paratroopers can go about their deadly business.

Seated across from me was Denis Barkats, a cosmologist who would stay on the Ice and by BICEP's side for the next eleven months in a job we refer to as "winter-over." Denis had done outstanding PhD thesis work at Princeton with Suzanne Staggs. Andrew Lange hired him in 2004, when my postdoc position ended and I left Caltech to become a professor at UC San Diego. I had wanted to hire Denis as my first postdoc, but, as it had been for me, the opportunity to work for one of cosmology's greatest luminaries was too tempting to turn down.

The flight to Antarctica gave us nine hours to chat, so I queried him why he was going so far away, for so long. Denis described how the chance to challenge himself, mentally and physically, appealed to him. But more than that was the chance to ensure BICEP was successful, and possibly answer mankind's oldest question: how did the universe begin?

He described the physical hurdles he had to pass to get to Antarctica. I was breathless just listening to him. Then he told me about the psychological tests he'd had to take to prove his winter-worthiness on the Ice: once fall starts, you're stuck there until spring arrives six months later. His test had several reasonable questions, such as, "True or False: My hands and feet are usually warm enough."[4] Then there were questions with somewhat inscrutable analytic intent, such as, "My sex life is satisfactory," and, "When I am with people I am bothered by hearing very strange things." Finally, there were some real head-scratchers, questions that made me doubt my own psychological fitness when I answered them in the affirmative: "I would like to be a singer" and "I have had no difficulty in starting or holding my bowel movement." It was all worth it, he assured me: because there was only going to be one sunset and one sunrise in the next year at the South Pole, he would take home $75,000 for a single night's work!

We enjoyed our conversation while seated in possibly the worst seats on the plane. We were at the end of one bench, meaning we were close to one of the honeypots. On our flight were two such buckets; one stand-up container for the men and a pail with a foam seat for the women, modestly surrounded by what looked like the actual shower curtain from the movie *Psycho*, complete with tattered edges and blood spatters.

The C-130 isn't really meant for long-haul passengers; it's more of a big, olive drab flying delivery van. A few feet away from me was Antarctica's entire supply of bananas—two thousand pounds' worth—which made for a delightful aroma . . . for the first thirty minutes of the flight. Positioning oneself for eleven hours amid such an olfactory palette is a Hobbesian dilemma—too close to the fruit and one is nauseated by their perfume, too far and the aroma of the honeypot overwhelms.

Nine hours after leaving Christchurch, and a week after I left San Diego, I arrived in Antarctica. I was finally at McMurdo Station, on the coast of the icy continent. Actually, I hadn't made landfall yet—the plane landed on a 3 m (10-foot) thick ice runway off the coast of a small volcanic island, Ross Island, where Robert Scott had established base camp nearly a century earlier. I was surprised by how warm it was; when I stepped off the 100,000 kg (250,000-pound) cargo plane, my feet sank into a puddle of water. Thankfully, the cargomaster had forced us to wear our sweat socks and vacuum-insulated boots.

A TOAST AT THE END OF THE WORLD

After a welcome briefing, I headed to one of the two bars on the base. Why two? Well, one was smoke-free, and the other wasn't. Unlike the non-smoking version, the smoking bar was also kept dark, adding to its allure and ensuring its popularity. In the hazy darkness, a few members of the BICEP team made a toast to the instrument that had brought us here, the one we hoped would take us farther back than any telescope before: "To BICEP, may it muscle its way back to the beginning of time!"

After one too many pints of Speight's Gold Medal Ale, I stumbled outside and caught my first glimpse of penguins playing by the water. They are wonderful creatures, as clumsy on land as toddlers, their little legs sliding out from underneath them when they try to walk, but as graceful as Olympic swimmer Michael Phelps in the water. The courageous determination of these flightless fowl emboldened me, and I decided to take an ill-advised hike up the 229 m (750-foot) high volcanic cinder cone known as Observation Hill. This was the hill used by Scott's party to survey weather conditions. It was also there that they watched for Scott and his team of four polar explorers to make their triumphant return from winning the race to the Pole. When they didn't return, his men erected a 2.5 m (9-foot) high wooden cross inscribed with the last line of Alfred Tennyson's poem "Ulysses": "To strive, to seek, to find, and not to yield." The haunting verse made for a restless night's sleep.

The next day I boarded a different C-130, this one kitted out with skis that did not retract. My flight to the Pole was only half full, with plenty of room to sleep or look out on the endless whiteness of the 600-mile-long Ross Ice Shelf. From the edge of the ice shelf it was 250 miles along the polar plateau to the Amundsen–Scott South Pole Station. By the time I arrived at the Pole, it was chilly for summer: –32°C (–25°F). I was also winded, for the Pole is more than 2.7 km (9,000 feet) above sea level. Three hours after leaving Ross Island, I checked into my room and collapsed, exhausted.

~

ON JANUARY 9, 1912, Scott and his team reached the most southerly point previously achieved, a mark he had set on his 1901 voyage. On virgin ground, he brimmed with confidence, the Pole assuredly within his grasp.

I also settled in easily. Though it was tedious work, toiling on a telescope in such an otherworldly location was incredible. I had to pinch myself as a reminder that it was really happening. The Dark Sector Lab (DSL) became a cozy home for BICEP, but it was unlike any building I'd ever been in before. Engineers had built the DSL on stilts, elevating it above the crystal desert plain, the way beachfront houses are built. In this case, blistering white snow, not sand, was the enemy. Without the DSL's spiderlike legs, the precious telescope on the roof would have eventually been engulfed under a horizontal avalanche—the fate of many buildings constructed in less prescient a fashion. The only building not on stilts (unsurprisingly) was the outhouse. A *functional* elevated outhouse would surely be worthy of a Nobel Prize!

The DSL was my fortress. It was an enclave of dim serenity, a hideaway from the Sun, sealed from the elements by 30 cm (1-foot) thick meat locker doors. In America, such massive doors are used to keep meat safely frozen; at the Pole, they kept the cold out and the astronomer "meat" fully thawed. Inside, lights flashed, fans whirred, and computers followed orders from unseen overlords, taking advantage of every second of precious summertime we had to get things working before we headed back north. We had a year's worth of work to do compressed into three months of daylight. The last plane left in mid-February, before temperatures dropped below the freezing point of jet fuel. If you missed it, you'd stay here for the longest night of your life.

We had so thoroughly tested BICEP that it was surprisingly simple to install it in its massive mount (fig. 46). After that task was completed, we spent most of our time seeing how well it was working, testing and calibrating it. Days blended into one another, each with the mocking Sun orbiting in a great and simple circle above our heads, never dropping below the horizon.

Life wasn't all tedium. There were many moments of downtime too. I passed my free time in the station's gym or on its basketball court shooting hoops with the grad students. I played ping-pong with plumbers and professors. We all talked about our plans for vacation after we left the Ice. I got plenty of advice for the camping

FIGURE 46. *Left*: Jamie Bock and the author installing BICEP in December 2007. *Right*: the "first light crew" (clockwise from top), Yuki Takahashi, the author, Jamie Bock, Denis Barkats, Ki Won Yoon, Cynthia Chiang. (BRIAN KEATING)

(known as "tramping" in native parlance) trip I'd planned around New Zealand's magical South Island. There was a real spirit of camaraderie at the station, as I imagine sailors on a ship must feel, which is fitting since it was the U.S. Navy that established America's permanent base at the South Pole, way back in 1956.

As Christmas 2005 (and, for the three of us Semites there, Hanukkah) approached, excitement grew. At the base, people made sure to celebrate. Before dinner there was the daunting-sounding "Race Around the World"—actually a 2.5-mile lap around the Pole that takes the runner through all the planet's time zones. I had no hope in this particular race—there were some serious athletes at the Pole, including one who had finished in the top 200 at the Boston marathon. Then there was a big lobster dinner for those who hadn't brought self-heating kosher meals. A few days before Christmas, I was analyzing fresh BICEP data. The whole system appeared to be working well, but there might still have been small effects lurking in the data that could insidiously mimic the B-mode signals we sought. We needed to tease them out of hiding and subtract them. We were looking for cosmic signals so minute that any instrumental systematic errors could completely overwhelm us. If we didn't fully

FIGURE 47. Scott (left) and his team at the South Pole, January 12, 1912.

understand the limits of our instrument, our environment, and even ourselves, failure would inevitably follow.

"Great God this is an awful place and terrible enough for us to have laboured to it without the reward of priority."

ROBERT FALCON SCOTT, JANUARY 17, 1912

ANTARCTICA IS UNFORGIVING of mistakes. One of Scott's biggest was his utter devotion to the scientific goals of the expedition. When he lost the Pole to Amundsen, he was depressed, but he could console himself with the prospect of the mission having achieved scientific success: the vast collection of Antarctic artifacts they collected. Yet even these samples contributed to his demise. Weighed down by sleds full of rocks, animal carcasses, and other specimens destined for London's Natural History Museum, the Englishmen moved even more slowly as the days passed. Though he lost the pole to Amundsen, who'd arrived nearly a month earlier, as Scott hoped, their artifacts did indeed make it to the Natural History Museum. But they bore a cost higher than any scientific expedition should suffer—they ended up dying for science.

My departure from the South Pole was, thankfully, less dramatic. Still, it was tinged with hints of foreboding and uncertainty. While calibrating BICEP just after Christmas, I received the most traumatic news I'd ever gotten: my older brother Kevin's phone call notifying me about my father's cancer diagnosis. He'd be undergoing chemotherapy treatments in Los Angeles in just a few weeks. Not only would I not get a chance to explore New Zealand, I wouldn't even be able to stay at the South Pole as long as I'd hoped. I scrambled to arrange the five flights that would take me from the South Pole to California. I needed to leave quickly.

My time on the Ice was so short I hadn't even had a chance to reflect on the project with Jamie Bock, co-leader and co-creator of BICEP. I wanted to know how he felt, what was going through his mind. I wanted to share our common accomplishment and reminisce about how a tennis court at Caltech had led us to all the way to the bottom of the world.

Jamie knew about my father's illness but didn't mention it to me as he escorted me from the station to my plane home. Instead of reminiscing, Jamie wanted to know if the rumors about me joining the other, competing, telescope called POLARBEAR were true. "I hear you and Adrian are working together. What's that about?"

POLARBEAR was an idea for a large telescope that had been conceived by Adrian Lee at UC Berkeley. Jamie and Adrian were about the same age and had both worked on similar projects in years past. Adrian had been a leader on the MAXIMA experiment, which barely lost out to Bock and Lange's BOOMERanG in measuring the universe's flatness.

In the highly competitive world of experimental astronomy, you often feel like you're only as good as your last telescope. You can never rest on your laurels, especially if you want to be a laureate of the Nobel kind. You must start thinking about your next project even before your current one is done—the universe has no shortage of mysteries, why should we stop preparing for them? If you don't, someone else will. So, you begin to network with other researchers, using your expertise on another project. Jamie and Adrian had long been competitors, and obviously Jamie felt betrayed by my joining up with Adrian.

POLARBEAR also would try to detect B-modes. But POLARBEAR was massive compared to BICEP. It was to be a reflecting telescope, more than ten times BICEP's size, designed to measure small angular scale signals that BICEP couldn't see. But POLARBEAR was at least four years behind BICEP. My group at UC San Diego had received a one-million-dollar donation to build the 3 m (10-foot) diameter POLARBEAR telescope, and it was obvious Jamie knew about it.

The fact that POLARBEAR was looking for a signal that BICEP couldn't see didn't dissuade Jamie. I was now more competitor than collaborator. I brushed it off: our CMB community is so small that almost everyone works on more than one experiment; conflicts are inevitable. Jamie himself was a lead contributor to Planck, building their polarization sensitive bolometers, which were identical to the detectors BICEP used.

He kept at it as we approached the hulking aircraft, its four massive engines whirling. The spinning props couldn't be stopped. And neither could Jamie.

"You've got to decide which team you're on," he yelled over the roar of the turbines.

I was confused, hurt, and angry. Was he seriously questioning my fealty to the experiment I had created? How could he do that now, with my dad fighting stage four cancer? My heart began to pound. I couldn't speak. The loyalty stemming from our days on Caltech's tennis courts five years earlier was gone; Jamie had transformed into a fierce competitor, with no love lost for his opponent. We stood in awkward silence for a few minutes as I tried to make sense of it all. The cargomaster yelled that it was time to go. I got on, breathless from a combination of altitude and anger.

On board, I was almost alone. A plane designed for fifty paratroopers now held just two Beakers. The crew let us go to the cockpit and I glimpsed the vast polar plateau stretching out before me. I had come so far, for so brief a stay, to leave under such disappointing circumstances. But BICEP's quest seemed trivial compared to the human drama that had played out so long ago. What must it have been like for Scott freezing to death on the icy plain below?

As the sunlight reflected off the ice beneath me, I reflected on what had happened in just a few short weeks. I was overcome with a range of emotions, from pride and satisfaction to anger, fear, and resentment. Most of all, I felt concern for my father. I had to make it back to L.A. before he started chemo. It had barely been a week since I'd learned of his predicament. I landed back in verdant Christchurch a day and a half after leaving the Pole.

It was pouring. I didn't even have a jacket.

TO ERR IS HUMAN, TO CALIBRATE DIVINE

After deploying BICEP, we felt we had a comfortable lead. For a while, no one was even attempting to beat us. With a monopoly, our spirits were riding high. Right from the start, BICEP appeared to be a success. We got many gigabytes of data back every day and the four genius graduate students who were getting their PhD theses working on BICEP (Evan Bierman, Cynthia Chiang, Ki Won Yoon, and Yuki Takahashi) analyzed the data just as fast as it poured in. BICEP was working well. We would have something to say about inflation soon—cosmically speaking, of course.

While we all knew about the false negative conclusion that had torpedoed poor Ed Ohm's prize, we were mostly worried about a false positive—seeing a signal that wasn't there, a pernicious B-mode impostor signal, one that mimicked inflation's imprimatur but didn't come from primordial gravitational waves. Unfortunately, there were many ways in which the sought-after signal might be mimicked.

As the BICEP data flooded in, my graduate student Evan Bierman searched for contamination that might be lurking within the Milky Way galaxy. He and I talked for hours about the physics of interstellar dust grains and the Milky Way's magnetic fields. It seemed hopelessly complex. The dust we were concerned about wasn't Hoyle's version—the whiskers he claimed scattered starlight and created the CMB. The existence of that dust, pervading all of space, between all galaxies and intervening matter, had been falsified by the DASI experiment's discovery of the minute level of the CMB's so-called "E-mode" polarization, the kind that traced density perturbations in the early universe, not the gravitational wave B-modes BICEP sought. The dust that concerned us was in our Milky Way itself. If its dust grains were slightly magnetic, then the Milky Way's all-pervasive magnetic fields could align them.[5] If things conspired just so, B-modes arising from the Milky Way's dust, rather than from inflationary gravitational waves, could mimic the B-mode polarization signal we hoped to see.

After a second season of observation was completed, Evan and I headed down to the South Pole to add new detectors into BICEP that operated exclusively at 220 GHz, where the polarized thermal emission from the Milky Way's dust would be much more obvious to spot; we could then remove it from BICEP's data. They were the first of their kind deployed to detect the Milky Way's imitator B-modes, and they helped us to understand dust's polarized emission, although only in the Milky Way's dusty disk. Evan's thesis demonstrated that cosmologists couldn't safely extrapolate polarization data from the Milky Way's disk to regions far away from the disk—and unfortunately, far from the disk was where we were looking for inflationary gravitational-wave B-modes.[6] But extrapolate was all we could do; no other polarization data existed.

Long before we published these results, I realized two things. As in the game of tennis that had led to BICEP, it was possible to commit both forced errors and unforced errors.

Unforced errors were numerous. Evan's analysis of BICEP's 220 GHz data showed that galactic dust could produce a large amount of spurious B-mode polarization masquerading as inflationary gravitational waves. What if we saw a B-mode signal, but had no idea what it was? To prevent a false positive unforced error, we needed to have sensitivity to the CMB's B-modes and the Milky Way's B-modes as well.

Nature forced a significant error upon us, by way of the inevitable random noise present in each of our data points. Noise from the hot glowing atmosphere of Earth, the detectors, and even the ice at the South Pole itself meant that BICEP could only detect inflationary B-modes if they were big—much bigger than signals which previous groups had already ruled out. We needed to lower the noise; we needed higher sensitivity to improve upon the limits set a few years earlier by NASA's Wilkinson Microwave Anisotropy Probe satellite.

To reduce the forced errors—in other words, to obtain more sensitivity—we had already begun the process of designing BICEP2, a polarimeter with five times more detectors, which would also be of a far more advanced design than the bolometers used in BICEP and Planck. The plan was to set a limit or, optimistically, to make a low signal-to-noise ratio detection (not what scientists usually want, but better than nothing) with BICEP, and then, bolstered by hints of a B-mode signal, go back and look at the same patch of sky with BICEP2, which would further reduce the noise.

So, in 2007, six years after my original pitch to Jamie and Andrew for BICEP, I instigated a new effort. By this time, I was a professor myself teaching physics and astronomy at UC San Diego. For most of my life, I didn't even know it was possible to be a professional astronomer and now I was one. (I had always thought being an astronomer was a mythological profession, like being a wizard. Who would pay me to do something so delightful?) No longer a postdoc, I was working for myself. I had a group of my own to worry about.

I proposed that my UC San Diego group and Andrew's group together build an upgraded version of BICEP, one whose array of detectors worked only at high frequencies. This high-frequency version of BICEP would be solely sensitive to the Milky Way's dust. It would be an insurance policy, just in case we saw something with BICEP—a way of ruling out a false positive claim of primordial B-modes.

This time, Andrew passed on my pitch, his objection being that my proposal "ran directly counter to the carefully thought-out logic of our proposal, which was to push as low on the B-modes as possible as quickly as possible," in a single band of microwave frequencies centered on 150 GHz.

Something had changed between us after I left Caltech and became a professor at UC San Diego, in 2004. Whether it was influence from Jamie, who clearly felt I had betrayed BICEP by joining POLARBEAR, or because Andrew wanted to nurture the careers of his current postdocs John Kovac and Chao-Lin Kuo, I'll never know.

For an anxious few days it even looked like I might have made one proposal too many—the group at Caltech had convinced Andrew that I wasn't needed on the BICEP2 team at all. Thankfully, soon after, Andrew changed his mind, informing me he was "sympathetic with your point of view, and less with my own. I think that you make good arguments for your rights to continued involvement in BICEP2, and I am eager to find a way for you to participate that everyone can be enthusiastic about."

I was okay with that. I didn't need to lead BICEP2; it was essentially an upgrade from the original. The primary change was the advanced superconducting detectors that allowed us to cram five times more detectors into the same precious, frigid thermos (fig. 48). "It's the original idea that matters," I comforted myself, figuring I'd get my fair share of the credit if we did succeed. But, in the meantime, we had to worry about not failing. To avoid being a modern-day Ed Ohm, this time ascribing a cosmological origin to

BICEP1
2006-08
98 detectors

BICEP2
2010-12
512 detectors

Each point shows
two detectors.

FIGURE 48. BICEP2 advanced detector array of superconducting transition-edge sensors (TES) fabricated photolithographically, resulting in robust, reproducible, mass-produced arrays that turned BICEP2 into a "microwave CCD camera." The tiling with planar arrays is extremely efficient: more than five times as many TES detectors were placed in the same focal plane area as in BICEP's semiconductor detector array. (NASA, © SHAFFER GRUBB)

a more pedestrian impostor signal, I couldn't dwell on hurt feelings or obsess over potential problems far off on the horizon. My group would have a big role in BICEP2, and my new graduate student, Jon Kaufman, and I needed to get to work.

Unlike BICEP, which had three frequency channels, BICEP2 only operated at a single frequency, 150 GHz, where the CMB is brightest. It would have ten times better sensitivity than the original BICEP but would otherwise be identical.

The philosophy was clear: first detect B-modes of any kind, and then determine whether they were cosmological, from inflationary gravitational waves, or more local, from the Milky Way. We called our strategy "B-modes or bust." BICEP2, like BICEP, was a Galilean refractor, housed within a cryogenic thermos. It would sit upon the same mount BICEP rested on, so we could continue to use the same South Pole site at the Dark Sector Lab, the most valuable cosmological real estate on the planet.

At the same time, it seemed that our monopoly on B-mode search efforts was coming to an end. The longer it took to build BICEP2, the more vulnerable we were to our competition, the billion-dollar Planck satellite. Planck launched in 2009, the same year BICEP2 was deployed. One million miles above Earth and its noisy atmosphere, Planck was the odds-on favorite to scoop us in our quest to capture the birth pangs of the Big Bang.

Because we had worked out all the kinks in the observatory and had much of the data analysis pipeline designed ahead of time, thanks to our three years of running BICEP, we could focus immediately on the performance of BICEP2's new feature: its superconducting detectors. What we saw at first got us all down—everyone, at least, except for our leader, Andrew Lange. He still had the confidence in us that we lacked.

THE ART OF COSMOLOGY

The Art of War is the classic treatise on military tactics and strategy. Although Sun Tzu wrote it for the Chinese military more than a millennium before Galileo was born, his tips for battle apply equally well to leaders of experimental cosmology campaigns today. Sun Tzu says a successful captain (principal investigator) must marshal "the army [collaboration] in its proper subdivisions [graduate students], the gradations of rank among the officers [postdocs, junior faculty], the maintenance of roads [detectors and telescopes] by which supplies [liquid helium] may reach the army, and the control of military expenditure [NSF funding]. These disciplines should be familiar to every general: he who knows them will be victorious; he who knows them not will fail."

Sun Tzu recommended mastering not only the logistical aspects of battle but also the intangible qualities of leadership: "The Moral Law causes the people to be in complete accord with their Captain, so that they will follow him regardless of their lives, undismayed by any danger. The Commander stands for the virtues of wisdom, sincerity, benevolence, courage and strictness."

Andrew was our moral leader. He guided by example, using his humanity as a tool to strip us of the cold competitiveness that sometimes bedeviled us. I knew it as soon as I met him, and in the decade during which I was the beneficiary of his academic paternal attention his kindness never diminished.

The summer before we were to deploy BICEP2 to the South Pole, I was married. Andrew and I had spoken many times in the year since he had rejected my dust-buster, the 220 GHz-only version of BICEP. I was hoping he would come to my wedding. Instead, he called me to say he was separated from his wife and would not be able to attend. He sounded so sad, so uncharacteristically down. I was crushed that my fatherly mentor wouldn't be there for me, but even more than that, I felt awful for his three sons. I knew firsthand how hard it is on young children when parents separate. I offered my deep sympathy and to be a patient ear to listen, if ever he needed one.

After I returned from the South Pole, when my dad was undergoing chemotherapy, Andrew called often just to check on me. He didn't have to; I was no longer his employee. He said to me, "Never lose sight of the big picture, your family. The more time you can spend with them, the better." He would make the time for me again and again, even when he was under tremendous stress as chairman of the Division of Physics, Math, and Astronomy at Caltech and dealing with myriad issues I knew nothing about. He was a gentle general, issuing orders from the heart.

It was therefore natural that Andrew would comfort all of us when, soon after BICEP2 was installed, the complex machine began acting in perplexing ways. The detectors were too noisy and we couldn't understand why. It should have been easy—we were using the same observatory, the same telescope design, and the same mount, identical in every way to the original BICEP experiment. BICEP2 even looked at the exact same patch of sky BICEP had stared at.

Andrew, ever the commander, leapt into action, battling not scientific forces but psychological ones. He called the team together and gave counsel as to what was going on with these new super-conducting sensors. Anticipating Sun Tzu's prescription, he began by imparting his wisdom, identifying subtle effects that might be causing the strange behavior we observed. Next, he implored us to keep looking, even if his hypothesis turned out not to be correct. He ended benevolently, cautioning us not to be too down on ourselves for not understanding all of BICEP2's quirks, and urging us to forge ahead courageously. He finished buoyantly: "We're learning how these detectors work!"—his trademark boyish optimism seemingly rekindled.

O CAPTAIN, MY CAPTAIN!

Eventually, things settled down on BICEP2. I had begun to split my time between POLARBEAR and BICEP activities.

Four weeks after BICEP2 began observing, on January 22, 2010, I was in the middle of a POLARBEAR collaboration meeting at UC Berkeley when Paul Richards, Andrew Lange's thesis advisor, burst into the conference room. Two decades earlier, he had supervised Andrew in that very room.

"Andrew is dead," Paul cried out. "He committed suicide."

I couldn't breathe. It was incomprehensible. The room fell silent. The first person I called was Jamie. He confirmed it: Andrew was gone. I went out into the cold, damp air and stared in silence for almost an hour.

~

A FEW DAYS AFTER HIS DEATH, Caltech physicist Sean Carroll created a memorial site on his blog, a place to share memories of Andrew, our larger-than-life leader, mentor, and friend. His friends, relatives, teachers, and even complete strangers bewailed the loss. They spoke of his boyish charm, his irrepressible humanity, and his brilliance. Most praised his mentorship, his guidance, and his leadership. From the expressions of grief and tributes on the blog I learned something new and something I knew. Andrew was once "Andy," a boyish, kind, generous young man. But he was also a seeker, and the rarest kind, for he had once been the first finder of universal truths.

BICEP2 had just been deployed. The stakes were so high. We needed Andrew more than ever. How could we go on without our captain?

~

A FEW YEARS LATER, I went back to where Andrew's remarkable life came to an end: a seedy motel, so utterly unworthy of containing the greatness of this sweet man. When I interviewed at Caltech a decade earlier, I had stayed in this very motel. The beginning of my life inextricably entwined with the end of his, at a crappy motel near the campus where he had once had it all: National Academy member, California Scientist of the Year, seemingly certain Nobel laureate. He had approved my inchoate ideas, giving me the confidence to pursue my dream experiment when it was a mere glimmer in my mind's eye.

I was deeply frustrated by questions that would forever be unanswered. Why didn't he reach out for help? Why hadn't he helped himself the way he helped so many others? How could he have done this to us? I never had the chance to thank Andrew for the whole panorama, the full exposure of my life that he had helped me develop. Nearly a decade later, I haven't found any words that suffice. What a loss . . . for me, for us, for his family, and for the cosmos.

Chapter 10

BROKEN LENS 2:
THE NOBEL PRIZE'S CASH PROBLEM

"There is no 'master plan' on the road to the Nobel Prize. It represents a lot of hard work, a passion for that work and . . . being in the right place at the right time."

AHMED ZEWAIL, WINNER OF THE 1999
NOBEL PRIZE IN CHEMISTRY

SAUL PERLMUTTER WAS IN A JAM. HE HAD LED the Supernova Cosmology Project for more than a decade and had little to show for it. It wasn't entirely his fault: he had to wait for a white dwarf in a binary star system to finish dinner. After gorging on its companion, the white dwarf would explode as a Type Ia supernova, temporarily outshining the galaxy it once inhabited. The cosmos wasn't on Perlmutter's schedule; it was taking years to acquire data from enough of these systems to go forward with his startling claim: that the universe was dominated by a mysterious form of pressure, dark energy, which was causing its rate of expansion to accelerate.

Eventually, after obtaining enough data, he and his colleagues (and his competitors on the High-Z Supernova Search Team) announced the discovery in 1998. And eventually Perlmutter won half of the Nobel Prize in Physics—in 2011, thirteen years later. His project was supported by the Department of Energy (DOE), which not only oversees the nation's nuclear weapons stockpiles, power plants, and green energy production but also is the dominant funder of physical science research.

During the 1990s, Perlmutter led his observing project from Lawrence Berkeley National Laboratory (LBNL), part of the DOE. His project was frequently reviewed by a committee, which often found that it didn't "fit the mission of the agency." According to Perlmutter, the limited amount of funding available and concerns over wasting it created a culture of fear. In conversation at the Times Higher Education World Academic Summit in 2016, he said, "You can be very good at not wasting money and also be very good at not making any discoveries."[1]

Perlmutter was lucky, he said. His funding was saved repeatedly, over the course of a decade, by a local division director of the LBNL. This director was authorized to make funding decisions locally and not required to obtain permission from "on high," meaning the DOE in Washington, DC. The director told Perlmutter, "You know, this project's been running for three years, hasn't found the very first supernova and yet it's worth staying with. And after six years it's still worth staying with and after nine years it's still worth staying with and then we can get results in the tenth year."[2]

Cosmological discoveries—indeed, most scientific discoveries—take time, but funding agencies must focus on the bottom line. Yet doing so exclusively may preclude the discoveries that justify more funding. As Perlmutter continued, at the Academic Summit, "Finally, when we started seeing the surprising results at the end, the [original DOE] committee said, 'This is exactly what we should be funding!'"

The blame is by no means the agencies' alone. Their broad mission statements—from building nuclear power plants to capturing light from supernovae—spread them thin. Scientists are partially culpable, too. The agencies need our support to serve on review panels to assist the program officers, to marshal support to enable sufficient resources to flow, and to get out of the lab to communicate to the public who support us. My fellow physicists should be inspired by Stephen Chu, Nobel laureate in physics who later became Secretary of Energy of the United States. If he can serve, so can we.

Sadly, funding agencies seem to be the only "strict constructionists" when it comes to Alfred Nobel's will—they want results and they want them soon. They are the most ardent adherents of Alfred's "preceding year" stipulation. According to Perlmutter, agencies now ask, "On what day will you make your discovery?"[3] He is convinced that his research project would not have won the Nobel Prize in today's funding climate; even if it had gotten started in the first place, it wouldn't have lasted long enough. His words are bracing. Demanding high return on investment *and* fast turnaround may be stymieing revolutionary discoveries in the future.

COMING HOME

On Monday, October 3, 2011, I returned to Providence, Rhode Island, to give the physics colloquium at Brown University, where I had received my PhD more than a decade earlier. It was almost two years since the deployment of BICEP2, and two years after Andrew's suicide.

The visit was exciting, the first time I had been back to Brown since my thesis defense. I had recently been tenured and won a few awards, including the National Science Foundation's highest award for young scientists, the Presidential Early Career Award for Scientists and Engineers (PECASE). I'd received the PECASE at the White House from President George W. Bush himself, for inventing BICEP.[4] And BICEP was what I planned to speak about back at Brown, my intellectual home, the singular place I identify with my becoming a scientist. The whole physics department seemed to take pride in my accomplishments.

I enjoyed seeing my former professors, including Nobel laureate Leon Cooper and Gerry Guralnik, who I expected would soon be a laureate as well. Gerry was the same brilliant mensch I remembered from a decade earlier, with wry sayings like, "To win a Nobel Prize you need good tools. The most important of these are public relations tools."

After the colloquium and a night spent with good food, wine, and conversation with my former professors now turned colleagues, I awoke to news that the Nobel Prize in Physics had been awarded to Saul Perlmutter (one half), Adam Riess (one quarter), and Brian Schmidt (one quarter), "for the discovery of the accelerating expansion of the universe through observations of distant supernovae." It was the first time the Nobel had been awarded to optical astronomers; all the previous astronomy-related Nobel physics awards had gone to theorists who calculated properties of astronomical objects, or observers who discovered phenomena, such as cosmic rays, or by using invisible radio waves or X-ray light, not optical light.

I was overcome with emotion. Only a few years earlier I had been at the Young Scholars Competition at UC Berkeley, a worldwide competition celebrating the birthday of Charlie Townes, winner of the 1964 Nobel Prize for the laser. The competition had been established to discover young scientists working on projects that would match the laser's beneficial impact on mankind. I had won first prize; Adam came in third. And while Adam's and my current research topics weren't related to lasers, I felt a twinge of jealousy wash over me. This wasn't supposed to happen. My proud, triumphant return to Brown degenerated into baser emotions. My brother Kevin provided the *mots justes* the way that only an older brother can: "You might have won the battle, but Riess won the war!"

Kevin was right, of course. Battle is an apt metaphor for what we scientists do. There is a fierce competition that begins the day you declare yourself a physics major. First, among your fellow undergraduates, you spar for top ranking in your class. This leads to the next battle: becoming a graduate student at a top school. Then, you toil for six to eight years to earn a postdoc job at another top school. And finally, you hope, comes a coveted faculty job, which can become permanent if you are privileged enough to get tenure. Along the way, the number of peers in your group diminishes by a factor of ten at each stage, from hundreds of undergraduates to just one faculty job becoming available every few years in your field. Then the competition *really* begins, for you compete against fellow gladiators honed in battle just as you are. You compete for the scarcest resource in science: money. Surprisingly, not by brains alone does science progress; funding is its true lifeblood. Cosmology's primary funding agency is the National Science Foundation. But the NSF proposal success rate is currently only about 20%, across all fields of physics and math: the lowest it has been in over a decade.[5]

CUI BONO?

Alfred Nobel was an idealist who wanted to reward those who "conferred the greatest benefit on mankind." Indeed, this selfless vision is etched into the physics prize medallion itself (plate 1). Above an image of Mother Nature is a line from Virgil's *Aeneid*: "And they who bettered life on earth by their newly found mastery." Yet Mother Nature herself, so red in tooth and claw, couldn't have devised a more efficient means of incentivizing bitter competition than the Nobel Prize. Indeed, the competition in science is at least as ferocious as in any corporate boardroom; there are many billion-dollar corporations, but the Nobel Prize is science's most closely held monopoly.

Most nonscientists think science is conducted by altruistic boffins, happy to find gainful employment doing work that they uniquely are capable of. Yet competition and science go hand in hand, and have done so since the invention of the scientific method itself. According to renowned sociologist of science Harriet Zuckerman: "Although differential ranking in science is not readily visible to lay observers [. . .] it is sharply graded." What is the cause of the stratification? Zuckerman claims that it's due, in part, to "differential recognition of scientists' contributions through citations to their work and honorific awards."[6] And, let's be honest, the Nobel Prize is the most honorific award there is.

Ralph Waldo Emerson once said, "Build a better mousetrap, and the world will beat a path to your door." This aphorism suggests that an ill-defined notion of excellence supersedes all other factors, even publicity. It may have been true in the nineteenth century, when Emerson and Nobel lived, but it hardly seems applicable today.

To benefit mankind, a brilliant discovery or invention is necessary, but not sufficient. To win the Nobel for your discovery, you must get there first. That, too, is insufficient. Most importantly, mankind must know about the benefit you have imparted; you must publicize your work, first within your field, then later to the broader base of scientists around the world. Upon hearing of your discovery, fellow scientists from within your field may nominate you for a Nobel Prize. Finally, a small subset of the Royal Swedish Academy of Sciences judges your discovery. If any one of these chain links should be broken, no Nobel Prize will hang around your neck.

Great art, too, can benefit mankind. But there is an important distinction between artistic innovation and scientific discovery. As historian of science Derek de Solla Price opined, "If Michelangelo or Beethoven had not existed, their works would have been replaced by quite different contributions. If Copernicus or Fermi had never existed, essentially the same contributions would have had to come from other people. There is, in fact, only one world to discover, and as each morsel of perception is achieved, the discoverer must be honored or forgotten."[7] Honor comes to those who do not wait.

Scientific discoveries often come cloaked in altruism. Take the golden plaque affixed to the *Eagle*, Apollo 11's Moon lander, which bears the signatures of the astronauts and their chief funding agent, President Richard Nixon (fig. 49). "We came in peace for all mankind," it reads. But the Moon landing was, at its heart, a battle in a war for global supremacy; the plaque was merely a piece of attire for a skirmish on the Cold War's most distant battlefield. When was the last time there was a battle to benefit mankind?

FIGURE 49. Replica of
the plaque attached to the
Apollo 11 lunar lander.
(NASA)

America's lunar residence lasted a mere three years, and we haven't been back since 1972. The Cold War endured almost two more decades. If lunar exploration was the path to world peace, it did not pan out. And if landing on the Moon truly benefited all mankind, how come the U.S. hasn't permanently occupied it? The ephemerality of the conquest echoed the race to reach the South Pole: once it was achieved, at great cost, the Pole lay fallow for forty years until 1956, when it became used as a permanent platform for science.

As with the lunar landing and the South Pole, there's no Nobel Prize for second place. The goal is to be the scientist who makes the first discovery, the one for the ages, the one who wins.

Winning the Nobel Prize confers the ultimate capital. I don't mean the nearly $1 million plus in prize money, nor am I referring to the gold medallion, worth another $24,000; rather, it's the intellectual capital which brings with it the power to set research priorities for entire scientific disciplines. In some cases, laureates even set the scientific agendas for entire countries. Most people think science shouldn't be competitive, but scientists are human, and humans love sport. And they cherish champions most of all.

Naturally, the Nobel Prize's biggest beneficiaries are the scientists themselves. While most scientists shun the spotlight, the Nobel Prize confers the "right kind" of fame. Laureates—the extreme

upper echelon of the scientific social strata—become the influenc-
ers, the trendsetters. Nobel laureate Paul Samuelson said, "Scientists
are as avaricious and competitive as [Adam Smith's] businessmen.
The coin they seek is not apples, nuts, and yachts; nor is it the coin
itself, or power as that term is ordinarily used. Scholars seek fame
[from] the other scientists whom they respect and whose respect
they strive for."[8]

After winning the ultimate accolade, laureates benefit from the
"rich get richer" phenomenon that historian and sociologist Robert
Merton called the Matthew effect, in which a greater proportion of
scientific resources becomes concentrated in the hands of a smaller
group of (mostly male) scientists.[9] Laureates receive resources
unavailable to their colleagues, and these come not only in the form
of research funding and lab space. Papers by Nobel Prize winners
garner more citations. Laureates attract the best graduate students
and postdocs. It's not that other great scientists can't attract funding, lab space, and graduate students—it's just that our society gives
laureates a gilded stamp of approval that makes them even more
desirable to funding agencies, universities, and prospective students.
And, since past Nobel Prize winners are automatically invited to
nominate future winners, their protégés receive the ultimate job
perk: they are far more likely to become laureates than those who
were not mentored by laureates.[10]

There's one last, if little-known perk: according to a recent study,
Nobel laureates enjoy an extra year of longevity compared to nominated scientists who didn't win.[11]

Next in the beneficiary list are the corporations and universities
that promote the Nobel Prize laureates on their rosters to sharehold-
ers and donors, respectively.[12] Funding agencies do the same.[13] Further down the list of beneficiaries is the Royal Swedish Academy of
Sciences and its Nobel Committee for Physics, which does a heroic
job vetting the nominations each year. While the Academy has
awarded a few prizes that haven't weathered well—such as the 1912
Nobel Prize in Physics awarded to Nils Gustaf Dalén "for his inven-

tion of automatic regulators for use in conjunction with gas accumulators for illuminating lighthouses and buoys"—they have been few and far between. But the Academy cannot be dispassionately detached from the prestige and influence its position affords. The awards ceremonies each year are a great source of Swedish national pride and are watched by millions. The ceremony itself has grown from a simple one-room affair in 1901, skipped entirely by Röntgen, the first prize winner in physics, to what promises to be a huge spectacle in the Nobel Foundation's glitzy new home in the center of downtown Stockholm, where ceremonies will take place starting in 2019.[14]

TARNISHED GOLD

Of course, the Nobel Prize isn't the only all-out competition that results in a gold medal for the winners. The first modern Olympic games were held in Athens in 1896, the year Alfred Nobel died. Like the Nobel Prize, they were endowed by visionary benefactors who hoped to head off the First World War, which was looming on the horizon. The modern games revived the ancient Greek tradition of the Olympic truce, during which youthful warriors laid down their swords and journeyed unmolested to Athens in hopes of encouraging "peaceful and diplomatic solutions to the conflicts around the world."[15] Unfortunately, the games, like the Nobel Prize, failed to realize those *fin-de-siècle* aspirations.

Olympic competition stresses athletes with pressures similar to those that dog scientists aspiring to win Nobel gold: grueling work done in isolation, over many years, for low wages. The costs required to train, travel, and compete to win an Olympic medal are astronomical: as high as seven million dollars per gold medal, according to a recent study.[16] Is there sufficient return on investment for national Olympic committees? The same could be asked of universities, where the financial packages used to lure Nobel laureates sometimes exceed Olympic medal amounts. Institutions, and the donors they must make proud, clearly feel the answer is yes.

There's one price tag for each Olympic gold medal that far exceeds that of Nobel gold. In the mid-1990s, the sports psychologist Robert Goldman posed "Goldman's dilemma" to elite athletes, asking if they would take a drug that guaranteed them a gold medal but would also kill them in five years. Approximately half of the elite athletes surveyed said they would take it.[17] One hopes that no young scientist would trade years of his or her life to win Nobel gold. But the pressures on young scientists are greater than at any time in the past. Many young scientists feel the ladder has been pulled up behind their senior colleagues. The reason for this, once again, comes down to the scarcity of resources.

A NOVELTY PENALTY?

In an ideal world, scientists could practice their craft without regard to politics. But in reality, big science takes big bucks. No university can support the large, experimental projects that are required to detect new forces and new particles. Only federal agencies like the Department of Energy, the National Institutes of Health, and the National Science Foundation have the resources. Sadly, federal funding for science has been in decline; it is at its lowest level since Eisenhower was president.[18] Just imagine trying to convince a U.S. senator (average age sixty-three) to spend a billion dollars listening for the echoes of black holes colliding a billion years ago while Medicare faces financial cuts today.

Now more than ever, funding agencies must justify their funding by trumpeting the success of the scientists they support. Hence, the prevalence of headlines touting the funding agency's role in prizes past and present.[19] The Nobel Prize confers the instant credibility that even a senator can understand. Certainly, young scientists are finding out that their funding comes at a great cost.

A recent article in the *Proceedings of the National Academy of*

Sciences written by Ronald Daniels, the president of Johns Hopkins University, found that the average age of a first-time recipient of a nationally funded grant has increased from under thirty-eight in 1980 to over forty-five as of 2013.[20] More disturbingly, the fraction of first-time recipients aged thirty-six or younger has plummeted from 18% in 1983 to 3% in 2010. The awful conclusion of the study is this: "Without their own funding, young researchers are prevented from starting their own laboratories, pursuing their own research, and advancing their own careers in academic science." With the success rate so low, the study continues, "it is not surprising that many of our youngest minds are choosing to leave their positions in academic research for careers in industry, other countries, or outside of science altogether."

In response to the low odds of getting funding, clever young scientists throttle back their proposed discovery potential, increasing their chances by proposing incremental projects that are likelier to succeed, rather than more revolutionary endeavors. Trading the ambitious for the incremental is far from optimal, according to Saul Perlmutter: "Simple step-by-step processes may occasionally be useful. But I think for the kinds of things you're asking world-class universities to be thinking about, that is not necessarily the best way to go."

Step-by-step processes are less novel by nature, and present fewer opportunities for serendipitous discovery. We should encourage bold new ideas, especially from young people. But funding agencies aren't as interested in rewarding novelty as they once were. A 2013 Harvard Business School study of medical grants confirms a disturbing trend: once projects go beyond a modest level of novelty, the probability they'll be funded decreases as their perceived novelty increases.[21] In other words, the chance of a proposal being funded has a peak, a maximum acceptable level of novelty. While this study concerned medical research proposal funding, I suspect the same goes for funding research in the physical sciences. Yet in physics, some of the most important and beneficial discoveries have been the most unexpected. How can we break out of this negative cycle?

We need to accept that big reaches will lead to big failures. A big part of society's risk aversion is how the public understands the process. The public largely sees science through its portrayal in the media. Right now, the media won't cover incremental science because it's not flashy, not front-page. They want the big stuff, because the public seems to care only about the big discoveries. This causes the media, driven by university publicity officers, to make more out of incremental discoveries than they may warrant.

Young scientists are often willing to take more risks than their older, more established colleagues. They are more resilient, and more able to endure the long duration between inspiration and recognition. Yet the two-year congressional budget cycle is punishing young scientists, and they are abandoning science, ultimately depriving mankind of the benefits of their bright young minds.

MODERN-DAY MEDICI

Fortunately, a new category of benefactors ready to support novel endeavors has appeared on the scene: private foundations free from the vicissitudes of short-term election-cycle competitive myopia. Even though they are far smaller than federal agencies, private foundations have begun to support research that benefits the public. They do this efficiently, for many reasons. First, private foundations have more autonomy; they are not beholden to governmental oversight with its fear of wasting limited federal resources. Second, foundations are less susceptible to the whims of review panels, which are often composed of competing scientists. Lastly, private foundations are immune to the highly politicized, two-year congressional budget cycle.

Foundations such as the Moore, Templeton, Simons, Kavli, Keck, Allen, Gates, Schmidt, and many others are changing the scientific funding landscape. These foundations are able to fund, and do fund, risky projects that require durations far longer than the typical public funding agency grant affords. They can nucleate, and then incubate, projects federal agencies are reluctant to support. About a third of the grants made by the Simons Foundation have been given to physical sciences and math, much of it earmarked "to support high-risk projects of exceptional promise and scientific importance."[22]

Over the next few years, a group of philanthropists called the Science Philanthropy Alliance intends to donate over $5 billion to basic science research, that is, research aimed at achieving fundamental understanding, rather than, say, producing new technology or cures.[23] While more than 80% of these funds are earmarked for the life sciences, that still leaves a substantial amount available for basic research in the Nobel-eligible physical sciences, including chemistry, physics, and astronomy—the last of which has great appeal to philanthropists.

Some have legitimately questioned whether the outsized influence of a small cadre of wealthy patrons is good for science. In the words of analyst Steve Edwards, they fear science is "becoming shaped less by national priorities or by peer-review groups and more by the particular preferences of individuals with huge amounts of money."[24] There are several responses to this concern. One is that even the enormous annual amount dispensed by private foundations to all branches of science—which might be over $500 million annually—is dwarfed by the National Institutes of Health's $33 billion annual budget.[25] This much smaller amount does not influence the NIH's areas of research; it merely supplements funding for other researchers.

Secondly, the biggest bang for the buck comes from the Nobel Prize. It has a far bigger influence on scientific priorities than do private philanthropists, many of whom fund prizes that are worth far more money than the Nobel. As physicist Lawrence Krauss pointed out in a *New York Times* op-ed, the new "big money," privately endowed science prizes have less luster than the Nobel despite their exceptional munificence: "I suspect that every single winner of the

Breakthrough Prize [an especially valuable award] would happily return the money in exchange for a Nobel Prize. The Breakthrough Prize, like the Kavli Prize and the other million-dollar-plus awards being given out around the world, will always be considered a consolation prize."[26] If prizes such as the Breakthrough Prize, worth nearly three times as much as the Nobel, do not have an impact on the healthy growth of physics, what would?

A wonderful use of the "big money" prizes would be to supplement funding for the experiments that ultimately allow for theoretical findings to enter the pantheon of "settled" science. Public–private partnerships can leverage the greater resources of the national agencies with the agility of private foundations to initiate risky ventures with revolutionary discovery potential.

In the past, private institutions like Bell Labs backed pure research that eventually led to innovations such as the transistor, the CCD image sensor, and the cell phone. Like Penzias and Wilson's discovery of the CMB, many of these discoveries were profoundly serendipitous. The scientist Tim Berners-Lee invented the World Wide Web not because of profit motivations, but because of a desire to share scientific information around the world.[27] Astronomy is replete with successful examples of public–private partnerships. In addition to the large investments in cosmic microwave background experiments by the Heising–Simons, Templeton, Keck, and Simons foundations, Paul Allen has given $30 million to create the Allen Telescope Array, which searches for extraterrestrial intelligence, George Mitchell has committed $25 million to the Giant Magellan Telescope under construction in Chile, and the Gordon and Betty Moore Foundation has donated over $200 million to fund one of the largest optical telescopes on Earth, the Thirty Meter Telescope.

Funding from these modern-day Medici is more crucial than ever. With their help, we might be able to welcome in a gilded age when scientists of all ages will once again confer the "greatest benefit to mankind" with their minds and the astounding new findings they discover.

THE SERENDIPITY TEST

The medal of the Royal Swedish Academy of Sciences, as the actual medallion that physics and chemistry laureates receive is known, is an artistic masterwork (plate 1). The obverse (front) depicts Alfred Nobel. Ironically, for an award which has only been given to two female physicists in its history, the reverse side features only female characters. Two women are set in an oddly compelling interaction: the Genius of Science is removing the veil covering the eyes of Mother Nature, who holds a cornucopia. The scene evokes the power of science to reveal truths to which mankind would otherwise be blind.

But sometimes scientists choose to be blind. Medical researchers, for example, guard against the ill effects of confirmation bias by using double-blind studies: both the subject under examination and the subject performing the test are shielded from information that might influence the outcome.

In astronomy, however, true double-blind experiments are impossible, since we cannot do true experiments. Some observatories, such as LIGO, use "blind injections" of fake events to partially immunize themselves from false positives.[28] These techniques do not perfectly shield the practitioners from events of which they should be, ostensibly, ignorant: almost as soon as LIGO captured its first binary black hole merger event rumors of the detection were leaked by team members, and immediately went viral on social media. If implementing double-blind protocols proves difficult in astronomy, what hope does cosmology have, where there is but one universe to study?

My proposal is that the Nobel Prize be used to alter this landscape by primarily rewarding serendipitous discoveries.

In his discussion at the Times Higher Education World Academic Summit, 2011 laureate Saul Perlmutter underscored the need for serendipity. "I think that there's a fundamental misunderstanding of what we are after when we are doing deep research of these kinds," he said. "I think that people forget that you are looking for gigantic surprises and transformations that allow us to do things that we never thought were possible and that you can't order them up in a certain way."

Of course, serendipitous discoveries can't be ordered up. But if serendipity were a prerequisite for the Nobel Prize, we could at least order up an environment that would incentivize novel discoveries that have true breakthrough potential. Furthermore, a prerequisite for serendipity would reduce confirmation bias—the subconscious tendency of researchers to look for information that conforms to their hypothesis and discard information that disagrees with it.[29] Bias is as bad for science as it is for society. Incentivizing discoveries which themselves are free of the corrupting influence of confirmation bias would go a long way toward increasing the transparency of the Nobel Prize process.

How would such a test play out, especially when serendipity is by definition unexpected? A serendipity criterion would have given Vera Rubin the Nobel Prize for the unexpected discovery of dark matter; the same criterion would have denied it to the experimentalists who looked for, and found, the Higgs boson. It would also remove from consideration the many experimentalists who are trying to detect dark matter, even if they succeed in detecting it, because they are directly trying to detect a specific target—as well as experiments like POLARBEAR or BICEP2, which searched for the signature of inflation.

Of course, my cosmology colleagues and experimentalists seeking to detect dark matter should absolutely *do* the experiments, receive other honors, and, most importantly, receive funding as well. But experimentalists should have a genteel adversarial relationship with theory. They should not be co-conspirators looking to "prove" a specific model, no matter which authority proposed it or how intellectually attractive it may be. Checks and balances are vital.

Serendipity has minted many Nobel Prizes, from Röntgen to the Curies, to Penzias and Wilson, to the 1974 winner Antony

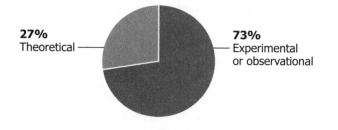

FIGURE 50. Percent of Nobel Prizes awarded to theorists versus experimentalists or astronomical observers. (© SHAFFER GRUBB)

Hewish and his non-laureate collaborator Jocelyn Bell. Valuing it specifically would redress the balance between Nobel Prizes given for theory and those given for experiment. As fig. 50 illustrates, to date three times as many Nobel Prizes have been awarded to experimentalists and observational astronomers as have been awarded for discoveries in theoretical physics. A serendipity criterion would mean Nobel Prizes would go to the theorist(s) who predict new phenomena, though they should win only *after* experimental verification.

The first Nobel Prizes interpreted Alfred's "greatest benefit" stipulation in favor of technology that bettered human life more or less immediately. Early prizes went to devices with medical applications, such as the X-ray and radioactivity, or practical applications such as Dalén's. Of course, some question the immediate benefit to mankind of today's theoretical awards. I agree: if the Higgs boson ever becomes relevant to your daily life, you've got much bigger problems than who should win the Nobel Prize!

While serendipitous theoretical discoveries may not provide immediately comprehensible—let alone tangibly beneficial—results, they typically do not require vast technical and financial resources. They can be made as easily—perhaps more easily—by the young than by the old. Making serendipity a criterion for the Nobel would help to keep our brightest minds working in the realm of physics, and level the playing field for those who are not already the recipients of funders' largesse.

DIVINE PROVIDENCE

Leaving Providence in October 2011, I knew it was too early for me to win the Nobel Prize. I had not yet had my revelation, my come-to-Alfred moment, that serendipity, rather than confirmation, should mint astronomical Nobel medallions. I was heartened about my prospects: the Nobel Prize for dark energy didn't go to the theorist who first conjectured it, Albert Einstein. He was long gone. It also didn't go to a theorist who had discovered a fundamental reason *why* dark energy had the value it had—that prediction is still elusive. Instead, the three astronomers (Perlmutter, Riess, and Schmidt) who observed the effect were the winners. There were half a dozen theorists who could lay claim to conceiving inflation—twice as many as Nobel Prizes available in a prize year. But there were only a few of us experimentalists who could *prove* inflation.

And so, after a close brush with Nobel glory, in a town whose name was a synonym for supernatural serendipity, I returned to California. I had no idea what fate would soon have in store for me.

Chapter 11

ELATION!

*"The most exciting phrase to hear in science,
the one that heralds new discoveries, is not
'Eureka!' but 'That's funny. . . .'"*

ISAAC ASIMOV

AFTER MY CLOSE CALL WITH THE NOBEL PRIZE at Brown, I left feeling energized and more determined than ever to win it. The praise lavished on me by my former professors and friends was flattering. But I wanted to earn science's ultimate accolade.

By 2011, BICEP2 was making great progress. We had learned a lot from its predecessor, BICEP, which had entered the final phase of data analysis. I still led BICEP, but not BICEP2, which was now, after Andrew's death, under the day-to-day control of Jamie Bock and John Kovac at Caltech. In their eyes, I was a competitor as well as a collaborator. Although I wasn't running BICEP2, I decided to take as active a role as possible. Especially important to me was supervising the PhD thesis of my graduate student Jon Kaufman, who deployed to the South Pole every year to help install, test, and analyze BICEP2 with the other team members. But we knew that it would probably be a while before high-quality data started to come in.

Getting funding, building a team, and going to the South Pole—all of that was hard. But at least it was somewhat within our control. The universe, however, was not on our schedule. We also had serious competition, most notably from a half dozen other ground and balloon telescopes and, of course, the Planck satellite, which had already set the tightest constraints on the inflationary gravitational-wave background using the CMB's temperature fluctuations. The same inflationary gravitational waves that would produce B-modes would also wrinkle space-time just enough to slightly distort the CMB's temperature pattern, but there was bedrock below that you couldn't see with the CMB's temperature alone. Beneath that level, which was seven times lower than the limit we had set with BICEP, we would be in *terra incognita*, exploring an unknown inflationary landscape with B-modes alone. Planck could explore that territory as well as, or maybe even better than, we could, according to studies the European Space Agency had released right after the satellite's launch in May 2009.[1] To make a significant dent, BICEP2 would need to improve upon BICEP's and Planck's results. And the improvement had better be dramatic.

In the end, the wait wasn't long at all, at least by cosmic standards. The signal was so strong it couldn't be true. In 2013, just three years after BICEP2 got first light and began its isolated sentry duty scanning the Southern Hole—which is what we called our patch of prime celestial real estate—BICEP2's own "that's funny" moment arrived. The B-mode signal coming in was *huge*. It was impossible to overlook. It was more like a crowbar than a needle, as BICEP2 co-leader, Clem Pryke, later put it. You'd think that would have made us happy. Instead, it made us very nervous.

WHO LEFT THE LENS CAP ON?

There were so many things that could have, even should have, gone wrong. We were trying to measure a temperature signal more than a billion times colder than the frigid South Pole. We didn't know if we'd ever see anything. Yet there it was.

Our first question was, "Who screwed up?" In the 1970s, Joe

Weber claimed to have detected gravitational waves in the lab.[2] He was a brilliant experimentalist and a co-inventor of the laser, who many thought had unfairly been deprived of his share of the 1964 Nobel Prize.[3] But Weber's waves seemed too good to be true—so good no one else could detect them. After his infamous blunder, we didn't want to be the second "discovery" of gravitational waves to go down in flames.

Such false findings happen surprisingly often in science. In 2011, a European experiment called OPERA made a startling claim: they had discovered neutrinos traveling faster than the speed of light. Since the speed of light—300,000 km (186,000 miles) per second—is as fast as anything can possibly go, this was an astounding announcement. Over a hundred years of theory and experiments had shown that the cosmic speed limit represented by the speed of light wasn't just a good idea; it was the law. The neutrinos arrived in the OPERA detector a mere sixty nanoseconds sooner than they should have.[4] Sixty nanoseconds might not sound like much, but it was more than enough to generate worldwide attention and Nobel rumors. The researchers' claim meant that Einstein himself was wrong, and it rocked the world of physics.

Eventually, OPERA's claim evaporated—the effect was traced to a loose cable that reduced the distance the neutrinos had traveled, explaining why they'd arrived at the OPERA detector prematurely. When the cable was tightened, the anomaly went away. But worldwide derision remained. Eventually, OPERA's principal investigators resigned from their leadership positions.[5] Between Weber and OPERA, we almost had more reason to fret over our findings than to celebrate.

And unlike the OPERA team or Joe Weber, we had serious competition: most notably, the Planck satellite, with its heavenly vantage point one million miles above Earth, free from gravity and atmospheric contamination alike. Planck possessed the perfect perch from which to scoop us, just as Penzias and Wilson had done to Dicke and company a half century before. Worse yet, the BICEP2 telescope had been disassembled a year earlier. We couldn't exactly go back and check to see if we had taken the lens cap off. But we could make use of our most powerful weapon: data, and lots of it.

We began by testing it for consistency by dividing the massive data set in half and making two maps, one from BICEP2's first eighteen months of observations and one from the second eighteen months. The two maps showed the same signal, albeit with lower signal-to-noise ratio (because each map had only half the amount of data as the two maps put together).

To prevent mistakes, carpenters say, "Measure twice, cut once." Well, BICEP2 astronomers cut the data dozens of ways, looking for discrepancies in data from one set of detectors versus another, or differences between when the telescope was scanning to the right versus to the left. We tortured the data in every conceivable way, each scientist on the team trying to concoct ever more outlandish scenarios that we had overlooked. Even if extraterrestrials had created our signal, the implications might have been less astonishing.

When I speak in public and am introduced as a cosmologist, I like to joke that you sure don't want me doing your hair and nails. Many people don't know that the similarity between cosmology and cosmetology is more than skin deep. They both have the prefix *cosm*, which is the Greek word for "adornment," as in the beautiful face the universe shows us. When I saw the BICEP2 data arranged into a map (fig. 51), the pattern of whorls and swirls took my breath away. It was exactly what inflation predicted we'd see, and it was love at first sight. The cosmos wasn't just beautiful. It was showing off.

Our exhilaration was mixed with a sense of foreboding. After a yearlong inquisition, it became clear: the signal was not coming from the South Pole, the atmosphere, nor BICEP2 itself. Where else could it be coming from, if not inflation?

One possible answer was that we'd seen the same material that had bedeviled so many astronomical discoveries since Galileo's time: dust.

Everyone knew that B-modes could come from interstellar dust in the Milky Way: microwaves scattering off dust within our own gal-

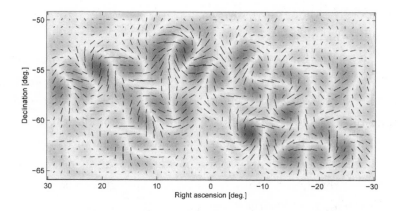

FIGURE 51. BICEP2's map of primordial B-mode polarization in the Southern Hole. (BICEP2 COLLABORATION)

axy could generate the pattern we saw. Might it make up the entire signal we were now seeing? How could we *prove* it was not dust, but the imprint of gravitational waves on the cosmic microwave background?

Though we had selected the Southern Hole—the patch of sky where BICEP2 hunted for B-modes—based on the low level of dust predicted by the best available models, we didn't know for sure if it was as free of contamination as we'd expected. What we really needed were high-frequency data.

Earlier I mentioned that the amount of polarization produced by dust increases very steeply with frequency. BICEP2 worked at 150 GHz only, corresponding to wavelengths of approximately two millimeters. Doubling the frequency would more than triple the dust signal. If dust were producing our B-modes, it would be obvious at 300 GHz . . . if only we had data at such high frequencies.

In truth, such a map *did* exist, one with the exact high-frequency data we needed. There was only one catch: it belonged to our competitor, the Planck satellite. And in early 2014, the Planck team hadn't yet released their B-mode polarization data. We were scared Planck might not only hold the key to proving our measurement right, but might have already glimpsed the inflationary B-mode signal before we did. If it really was as large as we thought it was, it was well within Planck's grasp.

We desperately tried to work with the Planck team, while being careful not to tip them off as to what we'd found. It was a perilous line to walk. By this point in the book you'll know that science teams that sometimes collaborate can be in competition at other times, particularly when there is a well-known goal or target signal both are looking for. This is a troublesome aspect of science; many of us treat the data as if it's "ours" when, in fact, it belongs to the people paying the bills: the taxpayers.

BICEP2 had much more sensitive data, but Planck's was broader, covering the whole sky and at many more frequencies than BICEP2 had. After everything else was ruled out, frequency coverage held the key to our fate.

The Planck team wouldn't cooperate.[6] Either they didn't have the data we wanted, or they did have it and they were going to scoop us. We had to go it alone. What BICEP2 lacked in frequency quality, we compensated for with quantity. We made five different models for the dust, each based on old data—the same data that we'd used to choose BICEP's observing region nearly a decade before.

Each of the five models predicted the total emission—the total heat produced by dust—at a particular region in the galaxy, but none of them could predict how much polarization we could expect in the Southern Hole. So, from these data, we extrapolated what galactic dust emission would look like in our patch if it were also slightly polarized. We played the guessing game, trying to be conservative, and eventually settled on a level of about 5% for our simulations.

Then came a revelation: we noticed that a Planck team member, Dr. Jean-Philippe Bernard, an expert on the Milky Way's polarization, had given a talk earlier that year which was posted online. Bernard showed an actual picture of Planck's dust measurements: a map of the sky as seen by our competition. It was a treasure map, with polarized "X"s marking the spot of sure Nobel gold.[7]

As soon as we discovered it, one of our team members digitized

Bernard's slide, revealing by extrapolation the formerly forbidden Planck data.[8] We knew it was an unorthodox approach. In fact, it didn't sit well with many of us. We took unpublished data, a single qualitative image, and digitized it, turning it into quantitative information. By doing so, we obtained a new model, one unavailable when we began taking data with BICEP, with exactly the information we craved.

Planck had not published this map and they likely had their own systematic errors to worry about. But the slide was public and freely available, giving us the green light to use it if we explained our methodology. But, if we went public, how much weight should this contraband slide carry? At first it was a curiosity, a digital trick to make us feel more confident. Then, a few months later, it snowballed, becoming a major link in the chain of reasoning assuring us that galactic dust was safely ignorable . . . and confirming something beyond our wildest hopes when we started: we had discovered B-modes from inflation.

Using the slide made me uncomfortable. On conference calls and in emails I complained to BICEP2's leaders. I wanted clarification: were we sure we had accurate measurements of dust? I was concerned that BICEP2's results had already been ruled out by Planck. Polarization of dust was the most obvious explanation for a signal we could see that Planck couldn't.

"How can we use slides that were shown in a talk but not intended for any quantitative purpose?" I asked in an email to the whole team. "My problem is that 5% number. How do we justify it to a referee/editor? Imagine if someone like Planck's principal investigator asks, 'Where did that number come from?' What will happen if it comes out that we got it from an unpublished talk slide?"

The leadership replied to my email, saying that it was fine to use the slide if we stated the assumptions we'd made. BICEP2's co-leader, Jamie Bock, was also a co-leader on Planck. He had taken over Andrew Lange's responsibilities on Planck, and his group at NASA's Jet Propulsion Lab had built the 353 GHz polarimeter channels on which Bernard's slide was based. Jamie must have seen the actual data and not just the slides, I rationalized. He knew that Planck could tell whether BICEP2 had seen the fingerprints of creation . . . or was the most sensitive dust detector ever made. While astronomers don't sign nondisclosure agreements, it's ethically questionable to use knowledge from one experiment to help another experiment, at least while both are ongoing. It's sort of like insider trading. But, come on, I thought, Jamie must have taken just one small peek at Planck's dust data and confirmed that it was safe to neglect. How could he not?

Plus, the Planck slide merely confirmed the results of the other five models we had, all of which showed that dust wasn't a plausible explanation for the bright B-modes we saw. Planck's slide would be but one piece of evidence, and not the most definitive piece of evidence at that. That distinction belonged to my precious BICEP, which had been renamed BICEP1.

Unlike BICEP2, which observed the sky at a single frequency—150 GHz, where the CMB is brightest—BICEP1 had three frequency channels, at 90, 150, and 220 GHz. With the benefit of these other frequency channels we could exclude, to some extent, the impact of dust above a certain level. When BICEP2 lead investigator John Kovac was asked by the journal *Nature*, "When did you first realize that you had detected the long-sought 'smoking gun for inflation'?" Kovac said, "Last fall, when we first compared the BICEP2 signal with BICEP1. That was very powerful because BICEP1 had very different detectors and used much older technology. So the fact that we were able to see the same signal with this completely different kind of telescope laid a lot of lingering doubts to rest. The remaining skeptics on our team were convinced at that point."[9]

I was one of those swayed skeptics. We could use Planck's slide, because it wasn't the main line of evidence. That most convincing evidence came courtesy of BICEP1, which said dust wasn't the cause of our signal, and we were 95% confident about that. In other words, dust had only one chance in twenty. Would you enter a lottery, the biggest one in cosmic history, if you had "only" a 95% chance of winning? Of course you would!

John Kovac made one last plea to the Planck team for their actual data, but again was denied. I figured Planck was about to scoop us. Waiting wasn't going to help. The Planck slide combined with BICEP1's data convinced all forty-nine of us, including me. I got off of my high horse. It was time: publish, or else our Nobel dreams might perish.

WHAT'S IN A NAME?

Publishing results is part of the way in which scientists stake their claims. Prior to the advent of scientific journals like the British journal *Nature* (established in 1869), scientists would perform experiments live, in front of, say, the entire Royal Society. Nowadays, not only does publishing a paper establish priority over the discovery, it also allows for the process of testing, vetting, and replication to begin. For BICEP2, there was one last pressing matter: what should we call the paper? You'll recall that Penzias and Wilson went with the modest "A Measurement of Excess Antenna Temperature at 4080 Mc/s," a rather understated title for a Nobel-worthy experiment.[10] For BICEP2, we wanted a headline that sold the B-mode beef with a little more sizzle.

There was plenty of Nobel precedent for us. Our title needed to capture the magnitude of the discovery the way that "Observational Evidence from Supernovae for an Accelerating Universe and a Cosmological Constant" had done for High-Z Supernova Search Team scientists Adam Riess and Brian Schmidt, winners of the 2011 Nobel Prize. While we went with a fairly measured headline, "BICEP2: Detection of B-mode Polarization at Degree Angular Scales," we were more aggressive in the text itself. That, too, was part of our strategy. We worried that if we didn't claim a detection of inflationary gravitational waves via the CMB's B-mode polarization—if we were too tentative and didn't declare we had seen the imprint of inflation—then Planck or one of our other dozen competitors could rightfully claim that *they* had discovered inflation. We didn't want to hit the ball out of the park and then forget to step on home plate.

The title was the final obstacle before going live in an unforgettable way. It became nearly impossible to keep the excitement contained. It was thrilling and I was bursting, but I was also sworn to secrecy. I was both exhausted and energized. How often does a scientist get to participate in science at the cutting edge of human understanding?

As we settled on a date for release of the BICEP2 discovery, I could hear the four principal investigators talking about the press conference during our weekly conference calls. Then rumors began swirling around Harvard University, where John Kovac was based, and where the press conference was set to take place. Just before the imminent media onslaught, the BICEP2 leaders made a website explaining what we had discovered.

Who leads BICEP2? *John Kovac has led the BICEP2 experiment. Clem Pryke has led the analysis that produced today's results. Jamie Bock contributed the optical concept for the experiment and developed the detector array technology. Chao-Lin Kuo designed the polarization sensitive detectors used in BICEP2.*[11]

The fact that I had created BICEP wasn't even mentioned. I had been its first champion, believing it could be done despite the many doubters who stood in the way. I had written the very first paper about BICEP more than a decade earlier.[12] Now I was blindsided. How could they? Was it because I had complained about using the Planck slide? Was it because I was co-leading the POLARBEAR experiment? Cynically, I thought maybe it was because only three people can win a Nobel Prize, and writing me out of B-mode history narrowed the pool by one. I wondered which of the other four co-leaders would get the ax next.

In the end, my role was relegated to a group acknowledgment in the website's closing credits:

Other major collaborating institutions for BICEP2 include UCSD, UBC, NIST, University of Toronto, Cardiff University, and Commissariat à l'Energie Atomique (CEA).

At the last minute, I asked Kovac for assurance that my foundational role in BICEP2 would be mentioned at the press conference. Begging for recognition from the man I had helped recruit to work on the experiment I had created more than a decade earlier was more than a little humiliating. He eventually agreed, saying the main reason I wasn't getting top billing was my role on the competing POLARBEAR experiment. But, I reminded him, Jamie Bock was leading our most competitive competitor, Planck. Kovac didn't respond. And, sadly for my role in cosmic history, the worst was yet to come. Reading farther down the website, I came upon this innocuous question.

What does "BICEP2" stand for? *Officially, "BICEP2" is not an acronym. It's simply a name.*

My heart sank. It was simply a name? BICEP2 followed on the heels of BICEP, which was the acronym I had proudly coined, an appellation for the first experiment ever to seek out the curling patterns of CMB B-mode polarization.

That stung. Names mean something. Children who have been adopted desperately long to know the name their birth parents called them by. And what is an astronomer? The word itself means "star namer." I knew the power of a name more than most; my surname had been changed when my stepfather adopted me. It was a reassignment of paternity then, and so it was now.

It would hurt to lose the Nobel Prize, but it hurt even more to have lost my brothers in arms. We had shared so much together, from the birth of our children to the tragic loss of our cherished friend Andrew Lange. We had suffered victories and defeats, sadness and joy. We had been together at the end of the Earth and now here we were, about to write the earliest history there is, and there was no ink left for me. I didn't understand why.

It felt awful. I didn't have to like it, but I had to accept it and move on. Instead of abandoning my hopes for receiving credit and slipping into anonymity, I decided to embrace the results and celebrate the cosmic fanfare of it all. I had helped write the Book of Cosmogenesis, In a century, it wouldn't matter how I was credited.

TREMBLING BEFORE CREATION

The week before the press conference, the media were alerted to an event to be held on Monday, March 17, 2014, at Harvard's Center for Astrophysics. No other details were given, only that the St. Patrick's Day announcement would herald "a major discovery." A handful of reporters, from outlets such as *Time*, *Nature*, and the *New York Times*, approved by Harvard's publicity officers, were told about the results.[13]

The night before the press conference, I couldn't sleep. On March 17, 2014, the secret was officially out. Harvard's press release read:

First Direct Evidence of Cosmic Inflation: Almost 14 billion years ago, the universe we inhabit burst into existence in an extraordinary event that initiated the Big Bang. In the first fleeting fraction of a second, the universe expanded exponentially, stretching far beyond the view of our best telescopes. All this, of course, was just theory. Researchers from the BICEP2 collaboration today announced the first direct evidence for this cosmic inflation. Their data also represent the first images of gravitational waves, or ripples in space-time. These waves have been described as the "first tremors of the Big Bang." Finally, the data confirm a deep connection between quantum mechanics and general relativity.

The press conference lasted for an hour. In the audience was physics royalty: Nobel laureate and CMB co-discoverer Robert Wilson was right in front of the presumed future laureates. Behind him were Alan Guth and Andrei Linde. Linde and Guth had been young men, separated by space and political ideology in the Soviet Union and the United States respectively, when they independently developed their cosmos-shattering revelations about inflation. Now they were together in triumph. It must have been a deep and satisfying vindication—the kind only experimental confirmation can provide.

In his remarks, John Kovac briefly mentioned my group at UC San Diego, but likely no one heard it—the world was erupting in applause. Cosmologist Marc Kamionkowski, one of the theorists who had inspired me to create BICEP, was most fulsome: "This is the greatest discovery of the century," he said. "If it sticks, which I think it will, it's Nobel-prize material."[14] MIT cosmologist Max Tegmark went even further, calling it "one of the most exciting moments in the history of science!"[15]

The funding agencies who had backed BICEP2 received their well-deserved share of the credit. "It reflects well on the NSF-supported BICEP team that they are now able to bring forward, for the first time, this long-awaited B-mode polarization signal so that their work can be examined by the broader astrophysics community."[16]

After the press conference, BICEP2's results were immediately posted on the open-source website arXiv.org. We were crowd-sourcing the peer-review process. Instead of restricting our findings to a single referee's eyes, which is what typically happens when scientists submit their findings to an academic journal—one who might well be a competitor and leak our results—we opened it to the whole world. Other experiments, like Planck, had done similarly, so we felt our temporary bypass of the peer review process had strong precedent. It was the beginning of the twenty-first century, after all, and high time the peer review process caught up.

As soon as the press conference ended, I heard news of a viral YouTube video. It was a gonzo-style production put out by Stanford University.[17] The video began like the opening of a Publishers Clearing House announcement of untold riches. Viewers tagged along with young Stanford professor and BICEP2 co-leader Chao-Lin Kuo as he walked up to the front door of Andrei Linde's home in Palo Alto.

Addressing the camera, Kuo says his mission is to announce that "BICEP has found the smoking-gun evidence of inflation." Then he coyly adds, "He has no idea that I'm coming!" Linde opens the door and Kuo breaks the news: "I have a surprise for you. It's five sigma at point two!" The meaning behind the cryptically quoted quantity is immediately apparent to Linde: BICEP2 had detected inflationary gravitational-wave B-modes at a stunning level of significance: the signal-to-noise ratio (SNR) is five, implying lower than one-in-three-million odds of being a statistical fluke. Linde is incredulous. He asks to hear it again. Kuo dutifully obliges. Tears well up as a bottle of champagne is opened. As it gushes forth, so does Linde: "We see the face of the Big Bang, it is an image of these gravitational waves, which is a purely quantum gravity feature of what was produced in the Big Bang. So, this is the remaining part of the story. If this is true, this is the moment of understanding of nature of such a magnitude that it just overwhelms. . . . Thank you so much for doing it for us."

All that was missing was an oversized novelty check made out for eight million Swedish kroner. Of course, Linde's gratitude was understandable. So much of his identity was tied up in inflation, it was impossible to see where "Linde" left off and "inflation" began. To receive confirmation within his lifetime was almost incomprehensible, a validation of the years spent in lonely, isolated combat against an invisible foe. For the BICEP2 collaboration, it was done less for Linde and more for ourselves; getting there "first," discovering something so obviously Nobel-worthy, was our reward. We, too, were lucky to experience success so quickly, compared to Linde. Nature had been kind, to all of us. The viral video brought tears to my eyes. I wasn't alone. It gathered two million views in a single day.

THE MICROWAVE MESSENGER

The BICEP2 telescope was as simple as could be, echoing Galileo's own. Delightfully, a refracting telescope had once again sharpened humanity's view of the cosmos and our place within it. Like a cosmic version of Galileo's self-published *Sidereus Nuncius*, the "Starry Messenger," the BICEP2 results paper posted on the Internet that day set off a media maelstrom. Overnight, scientists became celebrities and, gratifyingly, BICEP2 was the most famous character in the whole story.

The killer app for the universe had its IPO. BICEP2's stock catapulted into the stratosphere. It was a scientific unicorn, providing a billion dollars' worth of PR for its venture capitalists, which were, of course, the funding agencies that had backed it early on.

Like many early founders in Silicon Valley, where my initial ideas for BICEP were first sparked, I had been forced out. I felt like an astronomical version of Eduardo Saverin, the Facebook cofounder who sued Mark Zuckerberg after his ownership interest was radically diluted. For a nanosecond, I thought of suing BICEP2's principal investigators. But what grounds would I have besides sour grapes and spite? And what if Kovac and Bock were right: what if by joining POLARBEAR it was I who had betrayed them?

Although I had lost the best Nobel opportunity of my life, I found solace in having been there from the start. "The idea, that's the rub," I comforted myself, believing that was the prize that mattered most. I had come farther than I had ever dared imagine. By God, my idea had uncovered evidence for quantum gravity! How about that for a kid with a telescope!

Much to my delight, reporters restored BICEP2's acronym. The *New York Times* piece, "Space Ripples Reveal Big Bang's Smoking Gun," displayed BICEP's name in all its glory: Background Imaging of Cosmic Extragalactic Polarization. It felt good to have my founding role (partially) restored. But when my New York relatives read the *Times* story, they were astonished that the paper of record had no record of me whatsoever. All I could say was, "It was an oversight." That didn't satisfy my apoplectic mother. She demanded the *New York Times* issue a full-page retraction.

"If corroborated," the *Times* piece continued, the BICEP2 discovery would "stand as a landmark in science comparable to the recent discovery of dark energy pushing the universe apart, or of the Big Bang itself. It would open vast realms of time and space and energy to science and speculation."[18]

The speculation began immediately. Writing in *The New Yorker*, physicist Lawrence Krauss spoke of BICEP2's historic significance: "At rare moments in scientific history, a new window on the universe opens up that changes everything. Today was quite possibly such a day . . . if confirmed, it will have increased our empirical window on the origins of the universe by a margin comparable to the amount it has grown in all of the rest of human history."[19]

All that remained for the Nobel Prize was confirmation. It would be a nerve-racking wait, to be sure. It reminded me of how I felt each time my wife was pregnant: the waiting was hard but the present that came at the end made it all worth it. Surely Planck, or some other competitor we had left in the dust, would confirm our result. What mattered was that we had beaten them all to the punch. We had been first. Like the birth of the universe, the beginning mattered most of all.

Chapter 12

INFLATION AND
ITS DISCONTENTS

"When the Bank of Sweden established the prize for Economic Science in memory of Alfred Nobel (1968), there doubtless was—as there doubtless still remains—widespread skepticism among both scientists and the broader public about the appropriateness of treating economics as parallel to physics, chemistry, and medicine. These are regarded as 'exact sciences' in which objective, cumulative, definitive knowledge is possible. . . . Do not the social sciences, in which scholars are analyzing the behavior of themselves and their fellow men, who are in turn observing and reacting to what the scholars say, require fundamentally different methods of investigation than the physical and biological sciences? Should they not be judged by different criteria?"

MILTON FRIEDMAN, WINNER OF THE NOBEL MEMORIAL
PRIZE IN ECONOMIC SCIENCES, DECEMBER 13, 1976[1]

IN THE WEEKS FOLLOWING THE PRESS CONFER-ence, BICEP2 team members enjoyed the afterglow. There were rumors of a second Nobel Prize for inflation. A second one? It might surprise you to learn that years before BICEP2, there had already been a Nobel Prize awarded for inflation—albeit of a different kind: Milton Friedman had conjectured that the unemployment rate depended on the monetary inflation rate.

BICEP2 provided further evidence for Friedman's conjecture: our detection of cosmic inflation quickly garnered over a thousand citations, multiple media appearances, two TED talks, and big-ticket funding requests for follow-up polarimeters.[2] John Kovac was even voted one of *Time* magazine's "100 Most Influential People of 2014."[3] Friedman was right: by detecting inflation, BICEP2 ensured full employment for cosmologists for decades to come.

But what about Friedman's claim, quoted above, that "exact scientists" pursue "definitive knowledge"? Was there any truth to the assertion that the social sciences are more concerned than the physical sciences with the behavioral aspects of their practitioners?

The aftershocks of the BICEP2 press conference reverberated long after the camera crews had departed Cambridge. If inflation dominated the marketplace of ideas before BICEP2's broadcast, *after* our announcement inflation all but became a monopoly. The inflationary theory had eliminated all competitors, a situation that doesn't require a Nobel laureate like Milton Friedman to fret over. And, in truth, not everyone took delight in inflation's rising stock.

TINY BUBBLES IN THE MULTIVERSE

Inflation solved many of the Big Bang's problems. The two main questions had been why the universe was so flat, and why it was so uniform. But inflation had a serious flaw: the quantum field that drove the seemingly magical expansion—the inflaton—didn't come with an off switch. It kept inflating forever, and could not be quenched to allow the universe to gracefully exit and transition to the slower expansion that exists in the universe today. In Alan Guth's original model, the universe should be a pretty boring place: infinitely tenuous, expanding eternally, devoid of any matter, barren, featureless, and dull. We shouldn't even be here to complain about being bored.

Soon after Guth's "spectacular realization," Paul Steinhardt and his graduate student Andreas Albrecht rectified this flaw, finding a way to quench the inflaton field and stabilize it, at least under certain special circumstances. Widely separated regions of stability, forming a patchwork network, could each expand just enough to explain the universe's flatness and uniformity problems.

Independently, Andrei Linde came to the same conclusion as Steinhardt and Albrecht. By the early 1980s, Steinhardt, Linde, and another Russian cosmologist, Alexander Vilenkin, also realized that this new model of inflation would result in the universe eternally "self-reproducing" itself, spawning miniature bubble-universes separated from one another by enormous distances.[4] The multiplicity of universes became known as the multiverse.

Although I've decried the astronomer's dilemma—we can't actually do experiments on the stars themselves—we do have a few ways to work around this issue, at least by way of analogy. One of them is by borrowing tools used by our experimental biologist colleagues. If you place a few individual bacteria far apart in a Petri dish filled with agar gel (a bacterial delicacy) and let them reproduce, you can spawn your own microcosmic multiverse of sorts. The dish's agar gel is like the inflaton field, providing the fuel for the microbial island-universes to expand. Each bacteria colony grows in isolation, expanding in size on its own characteristic timescale. Eventually it fills the previously empty void with marvelous fractal structures, as shown in fig. 52.

While they lack the mathematical language to study their own properties, the microbial outposts exhibit long-range correlated behavior, which biophysicists describe using the mathematics of phase transitions—the same math behind cosmological inflation.[5] "Each colony is a superorganism, a multicellular organism with its own identity," wrote the late Eshel Ben-Jacob, professor of physics at Tel Aviv University.[6] Each culture behaves as if it is the only culture in the Petri dish, since nothing else affects it, at least for a while. Were they able, the microbes might proclaim: "The Petri dish is centered on us."

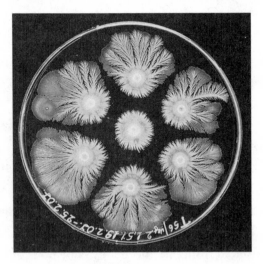

FIGURE 52.
Seven bacteria
universes in
the Petriverse.
(ESHEL BEN-JACOB)

Alone, the cultures would evolve, blissfully unaware of their actual utter banality. Countless generations would evolve until, at one moment, they might encounter a neighboring colony in the Petriverse. At that moment, they would realize they're neither unique, nor alone, nor in any way central to the vast expanse beyond their horizon. It would be the Copernican principle, writ small.

Linde's profligate cosmos had regions so isolated that each "pocket universe" could have its own unique physical properties, particle taxonomy, and even physical laws, completely distinct from and unrelated to the properties in another bubble. Linde's multiverse was a roiling sea of universes upon universes. It was a fractal pattern, one that was self-sustaining, eternal, and infinitely complex. Some cosmologists found the multiverse intoxicatingly beautiful; others found it abhorrent.

The multiverse and inflation quickly became synonymous. Alan Guth said, "It's hard to build models of inflation that don't lead to a multiverse . . . and evidence for inflation will be pushing us in the direction of taking [the idea of a] multiverse seriously." So handily did inflation spawn universes from nearly nothing that Guth declared inflation the "ultimate free lunch."[7] Linde agreed, adding,

"In most of the models of inflation, if inflation is there, then the multiverse is there. It's possible to invent models of inflation that do not allow a multiverse, but it's difficult. Every experiment that brings better credence to inflationary theory brings us much closer to hints that the multiverse is real."[8]

Even before BICEP2, many cosmologists searched for observable signatures of the multiverse.[9] But none was as convincing as detecting B-mode polarization. The multiverse would be all but proven if inflation was proven—and that's what BICEP2 had done. This, however, was what bothered some multiverse skeptics, such as Roger Penrose and even Paul Steinhardt. Not everyone was toasting inflation's triumph.

A PRESENT FOR NICOLAUS COPERNICUS

The multiverse had so much time, space, and energy to sprout new universes that it was inevitable there would be at least one bubble hospitable enough for cosmologists to exist in to speculate on such matters. This was not the first time such so-called anthropic arguments arose in cosmology, suggesting that observations of the universe must have physical properties that are compatible with the existence of the conscious life that observes it. Indeed, the multiverse concept and the anthropic principle were both first announced in Kraków, Poland, Nicolaus Copernicus's hometown, on the five-hundredth anniversary of his birth in 1973.

It was at this Kraków cosmology conference that physicist Brandon Carter coined the term "anthropic principle," while rejecting Bondi, Gold, and Hoyle's coopting of the Copernican principle as a feature in favor of the Steady State model.[10] Recall that, during the fourth Great Debate, these three Steady Statists had bolstered the case for their pet model by crafting the perfect cosmological principle, which extended Earth's banality from purely spatial, as Copernicus had claimed it to be in the first Great Debate, to temporal as well. In 1973, however, the Steady State was still very much alive, and many cosmologists, Carter included, were trying to kill it off for good. Doing so would require forgoing the Copernican principle, as Carter readily admitted: "what we can expect to observe must be restricted by the conditions necessary for our presence as observers. (Although our situation is not necessarily central, it is inevitably privileged to some extent.)" In other words, while the universe had no special locations or preferred epochs, it so happened that we lived at just the right time, in just the right place, to speculate about our ultimate centrality or lack thereof. What an ironic birthday present for Carter to give Copernicus, the first scientist to say we weren't at the center of it all!

To have enough time and space to find ourselves at a propitious time and place, Carter went on to create the multiverse, although he didn't call it that.[11] Carter himself recognized that the anthropic principle and the multiverse might be controversial, possibly even unpalatable, saying that even if attempts to derive the properties of the observed universe failed to find firm footing in a more fundamental mathematical structure, the multiverse idea "should still be taken seriously—even if one did not like it."

For some, Carr's present to Copernicus was a real disappointment. There was no reason these regions existed; it was just random chance, like saying the Earth needs to be within the so-called habitable zone—a precisely defined distance from the Sun such that liquid water can exist on its surface—because we inhabit it. Surely it would be better to have a reason *why* the Earth was located where it is, 150,000 million km (93 million miles) from the Sun. The tautological line of argumentation generated heated debate. Nobel laureate Steven Weinberg would later note that, if the multiverse is true, then "the hope of finding a rational explanation for the precise values of [the masses of subatomic particles] that we observe in our big bang is doomed, for their values would be an accident of the particular part of the multiverse in which we live."[12]

Whereas Carter's version of the multiverse was untethered to a physical theory, Linde's eternal inflation model practically mandated a multiverse. But it still had its fair share of anthropicity. An irrefutable, nonrational hypothesis was unscientific, according to its critics. Yet inflation was the best we had.

DENYING PATERNITY

The Steady State proponents performed a valuable role in the richness of the scientific debate. But with the demise of the Steady State came the end of the perfect cosmological principle. This opened the door fully to eternal inflation, the multiverse, and the anthropic principle—which is the fifth and final of the Great Debates I'll examine in this book.

As each of the Great Debates was waged, from Galileo and Copernicus versus the Church and the geocentrists to Hoyle and the Steady Statists versus the Big Bangers, the stakes rose. At first, merely the centrality of the Earth was debated. Later the very size, shape, and uniqueness of the Milky Way became the stakes. These first Debates were merely over humanity's *place* in the universe. There was no doubt that we inhabited *the* universe.

The fifth Great Debate pits those who say our universe is but one of an infinite number of universes against those who say that the multiverse hypothesis is not only wrong, it's actually bad for science. The main backers of the "Copernican side" of the fifth Great Debate are Roger Penrose and Paul Steinhardt. Penrose was the first physicist to point out that inflation required some very finely tuned conditions for it to work;[13] Steinhardt was the third of inflation's three fathers, the one least choked up in the delivery room at Harvard on March 17, 2014.

Years earlier, Steinhardt, with James Bardeen and Michael Turner, showed that the small fluctuations in the CMB could have been generated by quantum fluctuations during inflation. They predicted these fluctuations would grow to produce the large-scale temperature perturbations—and eventually cosmic structure like galaxies—later seen by the COBE DMR instrument in 1992. In 1993, along

with Robert Crittenden, Rick Davis, Dick Bond, and George Efstathiou, Steinhardt calculated the effects of gravitational waves on the
CMB's temperature pattern if inflation did indeed take place.[14] If
inflation could be proven, following in the tradition of the Higgs
boson Nobel Prize, Steinhardt would surely be booking his ticket to
Stockholm—with me, hopefully, sharing the smorgasbord.

But something fascinating happened in 2002, the year Steinhardt
shared the prestigious Dirac Prize for co-inventing inflation: he
began to condemn the theory he had sired.[15] The inevitability of the
multiverse, with its retinue of spawned universes ad infinitum, pre-
dicts essentially everything. To him, that was as good (or as bad) as
predicting nothing. "If you allow every conceivable possibility, then
there's no test or combination of tests that can disprove such a concept,"[16] he wrote. Why did Steinhardt want to disprove inflation? It
would seem more natural to want to prove it, as BICEP2 seemed to
do. But looming over him was the shadow of Karl Popper, a philosopher long seen as the defender of the scientific method.

POP(PER) GO THE BUBBLES

Just before BICEP2's announcement, Steinhardt wrote, "I think a
priority for theorists today is to determine if inflation . . . can be
saved from devolving into a Theory of Anything and, if not, seek
new ideas to replace them. This is because an unfalsifiable 'Theory
of Anything' creates unfair competition for real scientific theories.
Leaders in the field can play an important role by speaking out—
making it clear that *anything* is not acceptable—to encourage talented young scientists to rise up and meet the challenge."[17]

Steinhardt seemed to say: free lunches are nice, but sometimes
you get what you pay for. What triggered such opprobrium from one
of inflation's patriarchs? And what about Steinhardt's claim that the
unfairness of unfalsifiability is inherent in the theory of inflation?
What would that do to science itself? Would science ultimately lose
the ability to tell us anything about the actual origin of space and

time? Are theories that were not falsifiable not scientific? According to Karl Popper, the answer is yes.

Non-falsifiability has become a shibboleth for "bad science." Interestingly, though Popper is credited (or blamed) for proposing falsifiability as the criterion by which science is distinguished from non-science, it was never Popper's intent that it be so used in the "exact," or hard, sciences. Instead, Popper used falsifiability to assail disciplines that were non-scientific but were cloaked in scientific patinas so as to seem scientific. Fields such as astrology, Marxist dialectical materialism, and Freudian dream analysis drew Popper's ire.

For example, he wrote, "Astrology did not pass the test. Astrologers were greatly impressed, and misled, by what they believed to be confirming evidence—so much so that they were quite unimpressed by any unfavorable evidence. Moreover, by making their interpretations and prophecies sufficiently vague they were able to explain away anything that might have been a refutation of the theory had the theory and the prophecies been more precise. In order to escape falsification they destroyed the testability of their theory."[18]

One suspects Popper would be depressed to find that the falsifiability test has itself been falsified. More countries practice socialism than capitalism, exactly as Marx suggested. Dream interpreters are widely available in person and online. Worst of all, your local newspaper likely has far more space dedicated to astrology than to astronomy.

So where has the falsifiability test gotten us? Perhaps we are focusing too much on what can be falsified *now*, demanding immediate revelation of the limits of a new theory. Popper himself even agreed with this view.[19] Falsifiability is time-dependent. We shouldn't consider what is non-falsifiable today as being so forever.

According to historian of science Helge Kragh, even Popper himself didn't think falsifiability was science's *sine qua non*: "Popper did not assign any absolute value to the criterion of falsifiability and did not consider it a *definition* of science. . . . Far from elevating falsificationism to an inviolable principle, he suggested it is itself fallible and that it may be rational to keep even an admittedly wrong theory alive for some time."[20]

To Steinhardt, the anthropic reasoning underlying the multiverse was unscientific. Nothing about the actual properties of our universe could count as evidence that inflation was correct, for there were an infinite number of universes in the multiverse that weren't flat, weren't homogeneous, and didn't have the gravitational-wave B-modes BICEP2 had found. Wrote Steinhardt, "The multiverse idea is baroque, unnatural, untestable and, in the end, dangerous to science and society."[21]

EVERYTHING ETERNAL IS NEW AGAIN: THE GOSPEL ACCORDING TO PAUL

Like Hoyle, Steinhardt and his collaborators weren't content with merely attacking inflation. They constructed a rival model for the origin of the Big Bang: ekpyrotic (from the Greek for "conflagration") cosmology. In the modern version of this theory, which was first proposed in 2001, we refer to "bouncing" or "cyclic" cosmological models. These models use theoretic tools similar to those of the inflationary model, invoking new quantum fields that evolve in space and time. These models agree with current observations and, like inflation, they explain the large-scale properties of the universe that we observe today: flat and homogenous, but with small fluctuations such as those seen in maps of the CMB's temperature.[22] But the striking virtue of the bouncing and cyclic models is that, unlike inflation, they eliminate the "bang." The initial singularity is replaced by a "bounce," which is nowhere near as violent and avoids messy space-time singularities.

In the bouncing model, the smoothness and flatness of the universe occur over a long period of slow contraction as opposed to the brief, violent expansion of inflation. During contraction—before the bounce—quantum fluctuations that might nucleate island universes are dampened, not inflated. The bouncing and cyclic models do not spawn multiverses, and the un-falsifiable anthropic principle isn't needed.

Most importantly, the bouncing and cyclic models can be falsified: a virtue to those for whom falsifiability determines whether a theory is scientific or not. Neither the bouncing model nor the cyclic model produces primordial gravitational waves, so we would not expect to see any CMB B-modes. Therefore, the detection of B-modes was the definitive do-or-die test for these models—inflation's only competitors for scientific market domination. The marketplace of cosmic ideas was richer for the debate. But just as stock in these models was starting to rise, BICEP2's 2014 announcement sent it into freefall.

CHAMPAGNE IS SERVED AT THE ULTIMATE FREE LUNCH

MIT physicist Max Tegmark literally wrote the book on the multiverse.[23] Reporting live from the Harvard BICEP2 press conference, he said, "Today is a great day for most scientists except multiverse skeptics—at least in this particular universe. . . . Now it's harder for skeptics to dismiss this by saying 'inflation is just a theory': first they need to come up with another compelling explanation for BICEP2's gravitational waves. Today is also disappointing for the ekpyrotic/cyclic models that had emerged as the most popular alternative to inflation: they are ruled out by BICEP2's gravitational-wave detection."[24]

The irony must have delighted Andre Linde—the champagne in Stanford's viral BICEP2 video made all the more delicious courtesy of Linde's theory of bubble universes. Champagne, once uncorked, is notoriously hard to rebottle. The bubbles nucleate; they soar rapidly until they reach the surface where they explode, releasing the astringent aroma of bottled time and sunshine. It's a feast for the senses.

Linde's inflationary bubbles had been released thirty years earlier when he was a young man exploring the wonderful implications of his universe-spawning eternal inflation model. In the BICEP2 viral video, Linde mused on the importance of our findings: "Let's just hope that this is not a trick. I always live with this feeling: What if I'm tricked? What if I believe in this just because it's beautiful?"[25] Linde didn't have to "believe" in inflation any more than he had to "believe" in gravity: BICEP2 had provided evidence. Faith was superfluous. The beauty of his theory became yet more icing on the cake.

The day after the Harvard BICEP2 press conference, Linde, Kovac, and Guth gave talks about inflation and BICEP2 to a packed audience at MIT. Tegmark offered to teach Guth a little Swedish, for his presumed upcoming trip to Stockholm.[26] Guth then offered a toast over sparkling cider: "To the power of scientific reasoning!" he triumphantly proclaimed. The audience erupted in applause as, once again, tiny bubbles clamored toward the heavens.[27]

While BICEP2's detection of B-modes was a tour de force, seemingly confirming inflation beyond a reasonable doubt, for critics like Steinhardt it did little to winnow down the vast number of inflationary models that were consistent with our data. There wasn't just one model of inflation; there was literally an infinite number of them. Even the exact B-mode value we had detected did not allow scientists to discern which of the many types of inflation (chaotic, eternal, new, etc.) had produced it. "Some inflationary models can produce [a B-mode signal] as small as you like," said cosmologist Scott Dodelson of the University of Chicago.[28] Dodelson's comment was reminiscent of Hoyle's claim, upon hearing of the discovery of the CMB in 1965: "Had observation given 27 Kelvins instead of 2.7 Kelvins for the temperature, then 27 Kelvins would have been entered in the catalogue. Or 0.27 Kelvins. Or anything at all."[29] Now, with BICEP2, we had evidence for the multiverse, what Paul Steinhardt had once derisively called a Theory of Anything. Anything at all. . . .

Arthur Eddington once quipped, "Never believe an experiment until it has been confirmed by a theory." Eddington's dictum, of course, inverts the scientific method. Theory can no more prove an experiment than effect can precede cause. Days after our discovery, Steinhardt found a bit of gloom amid the brightness of BICEP2's discovery: "BICEP2 has brought in a lot of interesting debate about . . . what is the nature of science, this issue of whether it's important that science be testable or not testable, falsifiable or not falsifiable. . . . I've been hearing some very interesting views that . . . having a theory which is not falsifiable may be okay in science—and I find that very strange and actually I find it rather dangerous."[30] Later, Steinhardt grudgingly admitted that the pursuit of B-modes was "worthwhile nevertheless . . . because it is important for sorting out the correct scientifically meaningful theory."[31]

BICEP2 beatified inflation, entering it into the canon of astrophysical lore. Instead of contracting the field, as the discovery of the Higgs boson had, by ruling out vast regions of parameter space, our B-mode discovery expanded the theoretical landscape, and infinitely so.

The anthropic principle, too, was back on the table. And, thanks to the multiverse, the table had an infinite number of place settings. At his spot, Linde enjoyed champagne. Guth raised a toast with bubbly cider. For inflation's rival models, BICEP2 was a full glass of bitters with a falsification chaser.

Human knowledge has always evolved from the pseudoscience derided by Popper to actual science in fits and starts, sometimes enduring long periods of persistent unfalsifiability. Astrology— modern cosmology's intellectual forebear—had more to offer than just embarrassing stories of vengeful and capricious gods. Astrologers provided thousands of years of valuable observational data. They offered another service, too: they inspired astronomers to look to the heavens for evidence, not gods. There is a place for the unfalsifiable. Sometimes having an unassailable foe offers a useful foil against which superior theories can tilt, giving rise to theories far sharper than they would have been without competition.

BICEP2 granted inflation a stranglehold monopoly in the marketplace of cosmological ideas, stifling the fifth Great Debate just as it was getting interesting.

DO NOT PASS GÅ

This chapter began with Milton Friedman, who won the Nobel Prize for his work on economic inflation. It doesn't take a Nobel laureate to prove that monopolies are bad for society, undermining the consumer confidence that makes a marketplace thrive.

Days after the BICEP2 press conference, stock in the multiverse was also on the rise. Two of the three founding fathers of inflation seemed to be shoo-ins for inflation's Nobel Prize. The third patriarch, Paul Steinhardt's odds of going to Sweden seemed slim indeed.

But history would soon rhyme once again. In an ironic twist of fate, one even Hoyle himself couldn't have predicted, Steinhardt's alternative universe would receive its own bounce, courtesy of the most undervalued of all cosmic currencies: dust.

Chapter 13

BROKEN LENS 3:
THE NOBEL PRIZE'S
COLLABORATION PROBLEM

*"Do you not know that in a race all the
runners run, but only one gets the prize?
Run in such a way as to get the prize."*

1 CORINTHIANS 9:24

IMAGINE THE OUTCRY IF THE LEGENDARY 2016
U.S. men's swimming 4 x 200m Olympic freestyle relay team—
Michael Phelps, Ryan Lochte, Conor Dwyer, and Townley Haas—
obliterated their competition and came in first in their race, but only
Haas, Lochte, and Dwyer received medals, with nothing, not even a
silver, for Phelps. "Unfair!" you'd say. And you'd be right.

The Nobel committee seems not to recognize how collaborative science is today; their paradigm remains the lone genius, or a
duet or troika at most. Year after year, they perform their arbitrary
and often cruel calculus, leaving deserving physicists shivering in
the pool without any medal to show for it. Even those few modern
experimentalists who have won unshared Nobel Prizes owe their
success to numerous collaborators—especially in particle physics
and astronomy, which require massive data sets and large teams to
analyze them. No scientist gets to Stockholm alone.

The 2013 Nobel Prize in Physics, which was given to Peter Higgs and François Englert for the theoretical prediction of what was later called the Higgs boson, exemplifies four key problems in the selective awarding of the prize. First, it went to only two scientists (even though the committee allows three winners), when there were six other physicists, working in several teams, who independently introduced the idea and could rightfully claim joint custody of the Higgs mechanism. Higgs himself calls the it "the A–B–E–G–H–H–K–'tH mechanism," standing for Philip Anderson, Robert Brout, François Englert, Gerald Guralnik, Carl Richard Hagen, Peter Higgs, Tom Kibble, and Gerard 't Hooft.[1] All except Brout were still living in 2013.

Second, none of the over 6,200 experimentalists who helped make the detection at the Large Hadron Collider (LHC) will ever win a Nobel Prize. If the committee would even allow itself the indulgence of four laureates per prize, at least the two leaders of the ATLAS and CMS experiments at the LHC might have had a share.

In stark contrast, the 2017 Nobel Prize was awarded *only* to the instrumentalists who'd built the LIGO experiment. (Of course, the theorist who had predicted the existence of the gravitational waves that LIGO detected, Albert Einstein, had died sixty-two years earlier. Even I, who believe the Nobel should be awarded posthumously, think that's stretching it.)

Third, the award of the prize to Higgs and Englert blocked everyone else associated with the Higgs boson, whether experimentalists or theorists, from winning one. Even in clear-cut cases where historians agree that the Nobel committee made a mistake, never has more than one Nobel Prize been awarded per discovery or invention. Doing so would be tacit condemnation of earlier prize committees.

Fourth, the committee made clear that it prefers to confer no more than one Nobel Prize per person. (Only one laureate, John Bardeen, has won two Nobel Prizes in Physics.) So, since 't Hooft had already won a Nobel Prize in 1999 (for "elucidating the quantum structure of electroweak interactions"), the committee gave the 2013 prize to two first-time winners, despite the enormous role 't

Hooft had played. If the Nobel Prize is a true meritocracy, a scientist should be eligible to win it as many times as she or he makes a prize-worthy discovery. By that standard, Albert Einstein might have had as many as seven Nobel Prizes. That would certainly comport with his reputation among his fellow physicists.[2]

In truth, winning "only one" Nobel Prize isn't such an awful fate, even if it is shared. If the Nobel Prize were given to groups, the prestige of being a laureate would hardly be diminished; the fraction of the Nobel Prize a laureate receives is irrelevant, except in terms of the prize money (one-quarter of the total sum is the minimum amount a laureate can win). All winners receive the same 18-karat gold medal. Technically, Penzias and Wilson each won a quarter of the prize; the other half was awarded for completely unrelated work, as a sort of lifetime achievement award to Russian physicist Pyotr Leonidovich Kapitsa "for his basic inventions and discoveries in the area of low-temperature physics." Indeed, to the extent that fame is important, and I do believe it is, an educated layperson might know about Penzias and Wilson, but no layperson has heard of Kapitsa, even though he ended 1978 with twice as much cold hard Nobel cash as Penzias and Wilson did. And no one ever says, "Oh, Penzias, he only won a quarter of a Nobel Prize!"

TWO'S COMPANY

Alfred Nobel himself was an inventor, and he was used to filing patents to ensure that his claims were properly staked. When he wrote his will, in the late nineteenth century, science was done, if not strictly by loners, by single scientists with, at most, a handful of lab technicians. (They didn't have the students we professors rely on as our "force multipliers" today.) Had the Nobel Prize existed back then, Galileo would have won it in 1610, the year after he announced his serendipitous telescopic observations—and would not have shared it. No other invention, before or since—not the atom smasher, the X-ray, not even the automatic regulators used in conjunction with gas accumulators for illuminating lighthouses and buoys—had the transformative impact on physics, philosophy, and even theology that Galileo's telescope did; within weeks, it was clear that his telescopic observations had moved mankind away from the center of the universe. Copernicus, whose principle Galileo had verified, was long dead by 1610, rendering him ineligible. Hans Lippershey, widely credited with inventing the telescope, never actually observed the heavens with it, nor did his version have sufficient magnification to reveal the phases of Venus and the moons of Jupiter, which ultimately provided decisive evidence for the Copernican hypothesis.

It did not take long for the Royal Swedish Academy of Sciences to jettison the strict interpretation of Alfred's will. In the prize's second year, Hendrick Antoon Lorentz and Pieter Zeeman jointly won a sort of lifetime achievement award "in recognition of the extraordinary service they rendered by their researches into the influence of magnetism upon radiation phenomena." The prize was not given for a single discovery or invention (and, of course, their "service" hadn't happened in the previous year either). Following that, Henri Becquerel and Pierre and Marie Curie won for their work on radioactivity. In the two decades that followed, there were nineteen sole laureates. In contrast, as fig. 53 attests, the list of recent single laureates is small indeed. The last sole winner in physics was Georges Charpak in 1992.

It's still rare for more than a handful of theorists to discover a theory at the same time. By nature, theoretical discoveries are serendipitous and serendipity doesn't lend itself to multiples; three simultaneous lightning strikes are rare. Nowadays, it's much harder to be a sole laureate if you are an astronomical observer or an experimental physicist.

It wasn't always this way. Science was less collaborative in years past. More than twenty of the first thirty Nobel Prizes in Physics went to inventors or experimentalists, not theorists. The reason for this is shameful but, thankfully, since abolished: in the early 1900s,

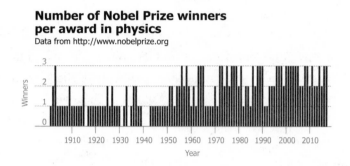

Number of Nobel Prize winners per award in physics

Data from http://www.nobelprize.org

FIGURE 53. Nobel Prizes in physics won by one, two, or three laureates. No single-laureate prize has been awarded since 1992, the longest such period in Nobel history. Gaps represent years with no Nobel Prize awarded. (© SHAFFER GRUBB)

European intellectuals derided theoretical investigations as anathema to physics, unworthy of Nobel consideration. The physicists who nominated laureates, some of whom were laureates themselves, considered pure theoretical investigations such as Albert Einstein's special relativity "Jewish physics."[3] Real physicists did experimental physics.

The movement away from lone laureates to multiple winners has accompanied near-inflation-like growth in all scientometrics—the metrics by which science, technology, and innovation are measured. Science historian Derek de Solla Price locates the inflection point in the "hockey stick" growth curve at World War II, when teams of physicists were kept "locked away in interacting seclusion. We gave them a foretaste of urgent collaboration in nuclear physics, and again in radar. By any metric, the image of the solitary researcher increasingly seems to become marginalized as a relic from the past."[4] This was the beginning of the period de Solla Price calls "Big Science," when research projects in all fields of science enjoyed exponential growth, creating a feedback loop that has taken us from entire fields with only a hundred researchers to single papers with ten times as many authors. We've gone from the Royal Society to the Large Hadron Collider in just over a century.

Today, the situation seems irreversible. While there is still diversity in the size of groups, many big projects with big goals require big telescopes and big dollar amounts. Biologist and philosopher Hub Zwart describes the ratchet-like behavior of Big Science as not only referring "to the actual number of researchers working and collaborating within a particular field, but also to the increased dependence of current research on massive, expensive and sophisticated technologies" such as LIGO or the LHC.[5]

With so many stakeholders, it shouldn't be surprising that the competition to win the Nobel Prize is extremely fierce. Of course, not all competition harms science; competition can also be healthy. It lends credibility to new discoveries: a signal detected by a single group doesn't mean that much without corroboration, and truly settled science becomes possible when more than one team gets the same results.[6] Multiple groups are needed to rule out mistakes and validate findings.

Yet excessive competition leads to wasted resources, the impetus (sometimes resisted, sometimes not) to publish prematurely, and a ruthless winner-take-all battle to get there first so as to capture the dwindling dollars from federal funding sources in decline. The size of new scientific projects, especially experimental ones such as large telescopes or particle accelerators, only makes the competition worse. Funding agencies are partially at fault for the climate of scientific competition, as Nobel laureate Saul Perlmutter, an outspoken critic of the current funding environment, has explained. Perlmutter's team, the Supernova Cosmology Project, was in a fierce battle with a rival team, the High-Z Supernova Team, to measure how the universe's expansion was slowing down over time. "They would race us to the results," Perlmutter has said. "Probably 90% of all the people on earth working on supernovae were involved in one of the two projects. It was a fiercely fought race. We wouldn't tell each other anything that was going on. We would be flying to the same telescopes they had just finished with."[7]

To their astonishment, the two teams independently found that the universe wasn't slowing down at all. Its expansion rate was, instead, speeding up. They had found evidence for dark energy, a mysterious form of anti-gravity—a latter-day version of inflation. Though they were in direct competition, members of both teams won the Nobel Prize.

In a study of the publication dynamics of Nobelists, science historian Harriet Zuckerman has found that laureates collaborate with more coworkers than a matched sample of non-laureate scientists. Yet, she observes, since the current rules compel the committee to overlook an increasing number of collaborating scientists, the award of the Nobel Prize to no more than three members often leads to the collaboration dissolving soon afterward.[8] Surely, this is not in the best interests of science.

I'd prefer that scientists be guided by the man who was the father of my field of observational cosmology. Bob Dicke declined Penzias's offer to be a third author on the Nobel Prize–winning CMB discovery paper, a decision which likely cost him a share of the 1978 Nobel Prize. While he may have lost out on science's top award, Dicke's group (at Princeton University) joined Penzias and Wilson's (at the private Bell Labs) to form a public–private partnership which allowed the Big Bang theory to achieve a wide acceptance.

VOTE EARLY, VOTE OFTEN

Early on, things weren't looking good for my chances at taking home Nobel gold. On BICEP2, I had forty-eight collaborators—seven times the number of theorists who predicted the Higgs boson—all of whom were alive in 2016 and either analyzed the data or worked on the experiment, or both. Plus, I'd been removed from the leadership role on the experiment that might have garnered the attention of prospective nominators. Lastly, as they say, "success has many fathers"; at least four theorists had conceived some version of inflation, already one too many for the allowed maximum number of laureates. Still, might I have had a chance to share in the glory, even though I'd been mostly hidden from view?

A few days after the press conference, I was still reeling, riding high from the magnitude of the discovery and a bit low from not having been at Harvard to partake in the festivities. On March 20, my spirits were buoyed when I found a poll on a popular blog run by physicist Philip Gibbs that said, "Vote for the Inflation Nobel Prize."[9] Of course, there was nothing scientific about this poll; it was more of a popularity contest than a true measure of the various scientists' contributions. Still, I took delight in it. More than 700 votes were cast, and when they were tallied, Andrei Linde came in first with 40%, followed not very closely by Alan Guth with 16%. Then came Alexei Starobinsky, a Russian cosmologist (10%) and Katsuhiko Sato, a Japanese physicist (7%). Both Starobinsky and Sato had come up with ideas like Guth's more or less contemporaneously. At number five was yours truly, with a whopping 4% of the vote, beating out fourteen other scientists who had greater than 1%.

The theorists got the most votes, a bias Gibbs attributed to their greater fame. So, according to this poll, I needed two of the four theorists to be displaced to make room for my Nobel Prize. Fortunately for me, at least, Gibbs said there was a way for the theorists to be kicked off: "The problem for these people is that no particular model for inflation has been shown to work yet. It is possible that that work has not yet been completed or that a more recent specific model will be shown to be right." Gibbs later wrote a journal article analyzing the results.[10] He had done his homework, and discovered the history behind my creation of BICEP. Reading on, I was even more heartened: "The experimenters are new stars so they have a smaller fan club and get less votes, but the Nobel committee may see it the other way round: If BICEP2 is confirmed by Planck then it will be clear that a Nobel worthy discovery has been made even if the theory behind it remains uncertain."

How could that be? There was a precedent for this type of experiment-only Nobel Prize in cases where the underlying theory remained unproven at the time of the award. In fact, it had happened three times in cosmic Nobel Prize history alone. In two of these, a CMB experimentalist won the Nobel Prize for observing phenomena predicted earlier by a theorist. On both occasions, the same man was left out: Ralph Alpher. Alpher was very much alive when both the 1978 and 2006 Nobel Prizes were awarded for the experimental discoveries related to the thermal relic of the Big Bang, the CMB, which Alpher (and Gamow, who died in 1968) had predicted.

The third observation-only precedent was also the most recent Nobel Prize in Physics awarded to cosmologists, the 2011 prize shared by Perlmutter, Riess, and Schmidt. As Gibbs said, "When the prize was given for accelerating cosmic expansion the committee made it clear that the award was for the observation *irrespective of how theorists interpreted it* and they are likely to see this [BICEP2] discovery the same way until it is clear that inflation is the correct explanation rather than the alternatives."[11] (Emphasis mine.)

I was heartened about my prospects: the Nobel Prize for vacuum energy didn't go to the theorist who first conjectured it, Albert Einstein (he was long gone), nor did it go to a theorist who had discovered a fundamental reason *why* dark energy had the value it had (that prediction is still elusive); instead, the three astronomers (Perlmutter, Riess, and Schmidt) who observed the effect were the winners—a precedent favoring the experimentalists. There were half a dozen theorists who could lay claim to conceiving inflation—twice as many as Nobel Prizes available for a prize year—but there were only a few of us experimentalists who could have been said to have detected inflation.

I was the highest vote-getter among the experimentalists in this not-so-scientific poll. If the Nobel committee felt similarly, I would have a fighting chance, even in a nineteen-way cage match. Things got even more exciting when the official Nobel podcast mentioned BICEP2: we were on their radar.[12] The *Guardian* rounded out the week after the discovery predicting a headache for the Nobel committee that year.[13] I would have been happy to supply the aspirin.

HOORAY FOR STOCKHOLM?

Recently, the most powerful scientific organization of its kind drastically changed the way it hands out its golden prizes. In 2009, the Academy of Motion Picture Arts and Sciences (yes, sciences) doubled the number of Best Picture Oscar nominees from five to ten, opening the wellsprings of credit to flow more fully.

Both the Nobel Prizes and the Academy Awards are meritocratic, determined by peers, and ostensibly egalitarian with no heed paid to commercial success. Both ceremonies are televised live, from giant halls filled with pomp and circumstance and guests in resplendent regalia. Winners receive golden idols from royalty, of the Swedish and Hollywood varieties. While Hollywood doesn't adhere to Alfred's stipulation that actors provide the "greatest benefit to mankind," there is a humanitarian award and a definite, if self-congratulatory, sense that the industry can influence society for the better.

When the physics Nobel Prize winners were announced in 2012, the physicist Jim Al-Khalili made several suggestions for modernizing the prize in an opinion piece in the *Guardian*. He piqued my attention when he wrote, "Most Nobel prizewinners will have carried out their breakthrough work for many years before they are recognized with the prize, and probably long after they had given up hope of that ultimate accolade—these are not the Oscars, after all, where an actor at least knows that he or she has made it to a shortlist. . . . For the rest of the scientific community around the world, this is also a time to hope that the winner comes from one's own particular area of research, boosting the chances of bathing in reflected glory and gaining valuable research funding."[14]

Al-Khalili's comment made me wonder: What if the Nobel committee recognized all the nominees each year? Currently, the names of the nominees (and nominators) are kept secret for fifty years. Why must the names of those who came close to winning the Nobel Prize be classified as if they were part of the Warren Commission report on the assassination of JFK?

The reason given by the Royal Swedish Academy of Sciences for the secrecy around nominations is to avoid upsetting nominees who do not win. This seems like a weak argument. Even though it's a cliché, Oscar also-rans often say, "It was an honor just to be nominated!" Announcing the nominees would redound to the benefit of the fields that are nominated as well. Scientists in those fields would receive more prominence and potentially more funding, just as those in the winner's field do, as Al-Khalili pointed out. It would also be gratifying, as a nominator, to know that your choice was considered. It's an honor to be a nominator, too, but if your nominee doesn't win, perhaps you won't waste your next opportunity on nominating the same person (assuming, of course, you haven't written a book critical of the Nobel Prize process and thus scuttled your chances of a follow-up invitation).

Of course, you might contend that revealing all the nominees could 1) take away from the winner's luster, and 2) put the attention on others instead of the winner. But, in practice, neither is likely to happen, just as it does not happen in the Academy Awards. Nobel Prize winners will always, rightfully, be recognized as society's fifth-degree black belt intellects. And, they will continue to rack up awards, seeing as they've already received the ultimate accolade.[15] In fact, they might appreciate some time out of the spotlight: the time demands on laureates are infamous, leading T. S. Eliot to opine, "The Nobel is a ticket to one's funeral. No one has ever done anything after he got it."

As aloof as scientists are toward celebrity, we could learn from our artistic counterparts. Hollywood's version of inflation is not of the cosmological sort, but regards recognition. According to a 2004 article in the *New York Times* entitled "Who Was That Food Stylist? Film Credits Roll On," the time it takes to roll the credits for a major Hollywood movie is pushing ten minutes[16]—triple the time it took the universe to make almost all of its hydrogen nuclei.

Modern Hollywood, like modern science, is more collaborative than ever. Fig. 54 shows the number of credited cast and crew mem-

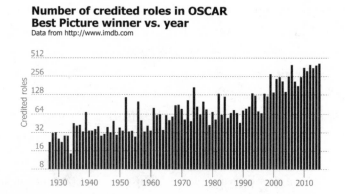

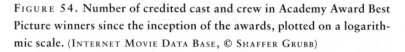

FIGURE 54. Number of credited cast and crew in Academy Award Best Picture winners since the inception of the awards, plotted on a logarithmic scale. (INTERNET MOVIE DATA BASE, © SHAFFER GRUBB)

bers (from director and starring actors to bird wrangler and on-set florist, as well as the vast teams creating computer-generated imagery) in Best Picture Oscar–winning films since the awards' inception in 1927. Compare this to fig. 55, which shows the number of credited collaborators on Nobel Prize–winning discovery papers since the prizes' inception in 1901. Both graphs show a characteristic "hockey stick" shape, increasing dramatically since the first Nobel Prize to Wilhelm Röntgen (one person) and the first Oscar for Best Picture to *Wings* (23 credited cast and crew), to 6,225 combined coauthors on the ATLAS and CMS experiments at the LHC and 353 credited contributors to 2014's Best Picture winner *12 Years a Slave.*

Hollywood has Stockholm beat when it comes to credit and awards; everyone involved with each year's Best Picture award receives a share of the credit. So too does each producer—Hollywood's version of the principal investigator—receive an Oscar for bringing the winning film to fruition. Artificially imposing a maximum of three laureates merely fosters unnecessary competition, and there's enough of that in science already.

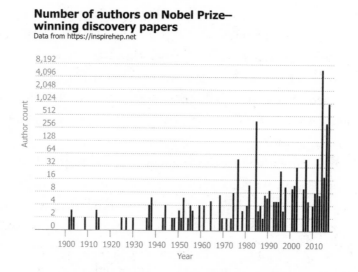

Number of authors on Nobel Prize–winning discovery papers
Data from https://inspirehep.net

FIGURE 55. Number of credited collaborators on Nobel Prize–winning experiments in physics, plotted on a logarithmic scale. Four particularly large values stand out: 385 authors on the discovery of the W and Z bosons in 1984, 6,225 authors on the two Higgs boson discovery papers in 2013, 342 authors on the neutrino oscillation discovery paper in 2015, and 1,004 authors on the LIGO gravitational-wave detection paper in 2016. Gaps represent years with no prize or prizes given for theoretical discoveries. (NOBEL PRIZE WEBSITE AND INSPIRE-HEP DATABASE, © SHAFFER GRUBB)

> "*Two are better than one because they have a good return for their work: if one falls down, his friend can help him up. But pity the one who falls down and has no one to help him up. Though one may be overpowered, two can defend themselves. Three are better even: Though one may be overpowered, two can defend themselves. A cord of three strands is not quickly broken.*"
>
> ECCLESIASTES 4:12

WHY NOT AN eight-fold cord, for the A–B–E–G–H–H–K–'tH mechanism? Or a thousand-stranded cord for all of those who collaborated on LIGO? As historian Elisabeth Crawford has pointed out, the original statutes of the Nobel Foundation didn't forbid awarding the physics prize to a group: "In cases where two or more persons shall have executed a work in conjunction, and that work be awarded a prize, such prize shall be presented to them jointly."[17]

Some have complained that giving a share in the physics prize to every scientist involved would devalue the award, decreasing the well-earned attention that the originators of the project deserve. Yet awarding the Nobel Peace Prize to groups has in no way decreased its prominence. The peace price can be awarded to groups, individuals, or groups *and* individuals (as was the case, for example, with the 2007 prize, half of which was awarded to the Intergovernmental Panel on Climate Change and the other half to former U.S. vice president Al Gore.

Especially in experimental science, where collaboration is essential, expanding recognition would help convince young people to take more risks in the ideas and projects they pursue. For me personally, the most rewarding aspect of my job is working with scientists from all over the world, from Uganda to the Ukraine, from Thailand to Texas, on every continent including Antarctica. It's high time the Nobel Prize reflects the true reality of modern physics: the best science of all is the most collaborative.

Chapter 14

DEFLATION

"The higher the target, the easier it is to shoot down."

OLD FIGHTER PILOT ADAGE

GALILEO HAD HIGH HOPES FOR HIS TELESCOPE. In addition to pure knowledge, baser desires impelled him. His first major invention—the military compass and its instruction manual—were reasonably profitable.[1] His telescopic discoveries made him a celebrity, which appealed to his grandiosity. It was natural that his design for the telescope would reap riches, too . . . if only he could maintain his monopoly.[2]

According to UC Davis science historian Mario Biagioli, Galileo's brilliant though somewhat sensationalist interpretations were at times in opposition to the far more staid "publication standards" of the day. No matter; the *Sidereus Nuncius* was self-published, which meant no peer review. And unlike his compass, the *Sidereus Nuncius* required no industrial production line, just a printing press. Whereas for each compass he had to provide a manual (unlike today's tech gadgets, which rarely come with such guides for the perplexed), the *Starry Messenger* became a manual without a gadget. The *Starry Messenger* was not merely Galileo's means of claiming credit for the instrument he developed; it was a way to sell his *ideas,* and do so at scale. He knew he could not make telescopes fast enough to satisfy demand, but he could sell the discovery and capitalize on the meteoric celebrity it brought. His ideas rapidly became dogma as well as currency, highly volatile lucre at that. Soon after, he secured patronage, and his fame spread rapidly throughout the European continent.

What Galileo's telescope failed to reveal, however, was that this notoriety would ultimately cause his downfall. His professorship in Padua brought Galileo closer to the Inquisition's Holy See, an all-seeing eye, one that had no desire to see the sights his telescope revealed.

In the end, Galileo's refractor never brought him much fortune. The applications he once hoped would bring income instead brought infamy. His works were on the Vatican's Index of banned books for centuries. After the press conference's lights faded, we and our refracting telescope would share his fate; BICEP2's newly famous scientists would face a withering inquisition of our own. Would our findings be vindicated, or would we too need a pardon, not from the Vatican, but from the entire scientific community?

MARCH 2014

EXACTLY ONE WEEK before BICEP2's St. Patrick's Day announcement, the POLARBEAR collaboration I was co-leading announced the discovery of B-mode polarization; not at the large, one-degree angular scales where BICEP2 had found them, but at small scales (six times smaller in angular size). POLARBEAR's pint-sized B-mode swirls were not from inflationary gravitational waves but instead from a nearer source. They were the result of gravitational lensing, a well-known phenomenon first predicted in Einstein's general theory of relativity nearly a century before: dark matter distorting the small-scale curvature of space (fig. 56).

POLARBEAR was a technological tour de force. But because no one doubted the existence of the kind of B-mode polarization signal POLARBEAR discovered, there was no fanfare: no media embargo, no press release, no press conference. Instead, as had become customary, we announced our findings by posting a preliminary version of the paper on the scientific pre-print server arXiv.org.

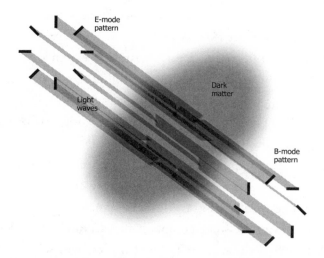

FIGURE 56. Gravitational lensing of the CMB's polarization can produce B-mode polarization at scales five times smaller than the gravitational-wave B-modes BICEP could resolve. (© SHAFFER GRUBB)

In the mid-1990s, B-modes were described by a young Argentinian physicist named Matias Zaldarriaga and his de facto advisor, the Slovenian physicist Uroš Seljak. Seljak and Zaldarriaga gave the signal its name. I had been friends with Matias when we were graduate students two decades earlier: he was at MIT when I was at Brown. Later, I watched his spectacular faculty career from NYU to Harvard and eventually to a permanent position at the Institute for Advanced Study (IAS), the Princeton institution essentially founded to give permanent employment to Albert Einstein. Matias was brilliant, inventive, and delightfully free of affectation, despite his permanent position at the storied IAS. On March 10, 2014, I emailed him about POLARBEAR's pre-print, the first detection of a signal he had predicted two decades earlier. "Congratulations," I said, "we've found your B-modes!" Matias replied immediately, "Yes, I saw. Congrats! Great to see the sensitivity has finally arrived." He then added, "I have been hearing rumors about primordial B-modes, even about press conferences etc. Can you add any information?"

Nope. For one thing, I wasn't privy to the details about the press conference. Soon, more theorists who knew about my role in BICEP sent me premature congratulatory messages, which were, in truth, thinly veiled requests for leaked BICEP2 findings. I was frustrated. With a week to go, BICEP2's discovery was the worst-kept secret in science. I didn't reply to Matias or any of the other theorists. They wouldn't have to wait long anyway.

A week later, on the day of the BICEP2 announcement, Matias again congratulated me. "Quite a long way from your grad student days! The results look awesome. I remember visiting Caltech when you showed me some computer figures of the first designs of BICEP1. I can't really believe it. The Universe was so generous and you guys did great. I am happy that it was you guys. Only sad note is that Andrew Lange is not around." It was so gratifying to receive praise and recognition from such an eminent scientist, writing to me all the way from Einstein Lane!

APRIL 2014

BASKING IN PRAISE did not last long, though. It was time to get back to work. We still had to submit the BICEP2 paper to the *Physical Review* for peer review. Within three weeks of the press conference, 250 scientific papers had been written about our results. That was astonishing; a paper is considered "famous" if it has 250 citations over the course of decades! There was even a paper suggesting that BICEP2's data might contain some contamination from the Milky Way; a group led by Oxford's Subir Sarkar proposed that synchrotron radiation, a well-known emission from the Milky Way galaxy and the type of radiation Penzias and Wilson were originally going to study—before the pesky CMB sidetracked them—might be the source of BICEP2's signal.[3] BICEP2 scientists pounced on Sarkar's criticisms and declared them unfounded, inconsistent with both BICEP2 and the earlier BICEP1 data. It seemed that some theorists wanted to benefit from the attention BICEP2 had gotten, if only in reflection. In any case, as Winston Churchill said, "Nothing in life is so exhilarating as to be shot at without result."

Then, in early April, I got another email from Matias. How many times can he be congratulating me, I wondered?

"*When the dust is low, but spread over a wide area, it betokens the approach of infantry.*"

Sun Tzu, *The Art of War*

MATIAS'S APRIL EMAIL was no "attaboy." He was disturbed. He wanted to talk details. What did I know and when did I know it? It was the beginning of a trial I had long feared. Rumors were swirling at Princeton about the way we had used the infamous Planck slide. "People here in Princeton are very concerned about dust," he said, ominously adding, "In fact they have managed to convince me that there is not a very good reason for me to believe it is not just dust. Have you looked into the foregrounds yourself?" Of course I had looked at the foregrounds—potential sources of contamination such as polarized emission from the Milky Way's dust. The whole team had been worried about our galaxy producing spurious B-mode polarization that would masquerade as primordial gravitational-wave B-modes. But data at low frequencies from BICEP1 and at high frequencies from Planck's scrubbed PowerPoint slide convinced us we were okay.

A few days later, I got wind of a colloquium that Princeton University's David Spergel had given just after the Harvard press conference.[4] David said he had spotted a blunder in our results, that our data were contaminated by dust within the Milky Way galaxy. Soon, I found out there were others at Princeton laser-focused on the way we modeled dust. The BICEP2 leadership had anticipated an onslaught, perhaps even a backlash, from the Princeton folks, who were working on several competing B-mode experiments. Maybe they were just frustrated after being scooped on another major

CMB discovery. Like Sarkar's synchrotron hypothesis, I suspected we'd soon rule out more Milky Way mimicry as well. But the attacks kept coming.

I asked Matias if it was David Spergel alone causing his concerns. Ominously, Matias said, "I think there is nothing else people here talk about." My heart stopped. Princeton's cosmology program is the top-ranked in the country—cosmology's own Holy See, comprised of the world's best experimentalists and theorists, among them multiple members of the National Academies of Sciences. It felt like an inflationary Inquisition, one that could put the BICEP2 results on a modern-day Index of banned pre-prints.

Imagine finding out the entire IRS is obsessed with *your* tax return. Not just one rogue auditor, but everyone, from the Secretary of the Treasury on down, fixated on your Form 1040! It was petrifying.

Matias told me that an outstanding young physicist named Raphael Flauger was leading a paper with Spergel and Spergel's graduate student J. Colin Hill. Flauger had convinced Matias that the Milky Way's dust polarization was higher than what the BICEP2 scientists had assumed. We were vulnerable to the same sort of tactics we had employed in utilizing the unpublished Planck slide; they could digitize our results before we released them. Live by the slide, die by the slide.

Matias added, "Don't get me wrong. Obviously, there is nothing more I would want than the result to be correct. But the discussions here have shaken my confidence and thus I hope you guys respond to the skeptics with a detailed explanation of exactly what you did with those Planck slides."

MAY 2014

BY EARLY MAY, Flauger and his collaborators had finished their analysis, and it didn't look good for BICEP2.[5] According to Flauger, we had used an incorrect estimate of the level of dust polarization in the Planck slide, a value four times lower than we should have used. If true, BICEP2 would go down as the most celebrated dust detector in history—tricked, like so many before us, by a dirty mirage.

But Flauger's analysis wasn't conclusive. He himself remained dispassionate, saying, "I hope there still is a signal. I'm not trying to pick a fight; this is how science works, that someone presents a result and someone else checks that. But it doesn't usually happen in public like this."[6] He and his colleagues, as well as Uroš Seljak and Michael Mortonson, claimed our interpretation of Planck's results was suspicious; but this didn't mean we were wrong.[7] Only new data, data unavailable to either BICEP2 or the groups doing the reanalysis, could tell us if the ax would eventually fall. The jury was still out.

Flauger's analysis was thorough, and it took several weeks for other cosmologists to digest it. A tense atmosphere settled over the CMB community; this was a cosmic cliffhanger, slowplaying us all. Then, on May 30, 2014, a Nobel prizefight took place in New York City in the form of a debate on inflation held at the 2014 World Science Festival, moderated by physicist Brian Greene. Next to Greene were three of inflation's fathers: Paul Steinhardt, Alan Guth, and Andrei Linde.

On Paul's left sat Professor Amber Miller, a CMB experimentalist and Dean of Science at Columbia University. On her left was BICEP2 principal investigator John Kovac. In front of the live audience, Greene transformed into a fight umpire as Kovac squared off against the inflation patriarch turned apostate, cosmologist Steinhardt: the young Harvard professor vs. the old guard advocating that the field needed more adult supervision.

Linde and Guth acquitted themselves adroitly, but both seemed mere spectators to the two pugilists trading B-mode body blows. Kovac deftly bobbed and weaved, but Steinhardt landed a haymaker, asking Kovac if he had backed away from his claim that a "new era of B-mode cosmology has begun." "Do you stand by that still?" Amber Miller sat uncomfortably, arms crossed, jaw set, in case Kovac and

Steinhardt came to actual blows. Kovac, seemingly on the ropes, holding on for the final bell's salvation, countered with, "We said *apparently* over!" Mercifully, the event concluded soon after, but BICEP2 was badly bruised—a technical knockout for now.

The beginning of the summer found the BICEP2 team in full panic mode, analyzing and reanalyzing data, responding to referee reports and putting out fires in the media and at scientific conferences. Paralleling our scientific battles was a battle *in* the media *about* the media. In particular, the propriety of the Harvard press conference became one of the hottest topics in all of science. The criticism we received about the way BICEP2 sought publicity was almost as intense as the heat we took for using the scrubbed Planck PowerPoint slide.

STOP THE PRESSES!

Scientists, pundits, and journalists alike questioned the decision to announce our findings at a press conference before peer review had been completed. While it's impossible to know whether holding a press conference was good or bad for us specifically, the issue of if, and when, press conferences should be held is an important one. Such decisions are always stressful. For a physicist, a press conference is likely a once-in-a-lifetime event. If your results are correct, a press conference–worthy discovery might result in a Nobel Prize. If your result is erroneous, it might be the end of your research . . . and its press coverage.

For BICEP2, the standard practice—a months-long peer review process, which would then be followed by a press release—had many disadvantages, any of which, individually, were worrisome. In total, they were completely unpalatable. First off, during the peer review, rework, and resubmission cycle we could have been scooped by the competition. Second, we feared that sending the paper to a journal would be unfair, giving a particular group—referees and their friends—a head start on proposal submission. My field is so competitive that the only people who weren't on BICEP2 who could have reviewed the highly technical aspects of the paper were competitors.

Our first priority was to make a scientific presentation to communicate our results to all our peers in the cosmology community. By releasing BICEP2's papers and data online, we allowed the entire community, not just two referees, to immediately begin a technical review. While some scientists praised our decision to go public first, analogizing our decision to the announcement of a blockbuster new drug, the criticism of BICEP2's crowd-source approach was, at times, brutal. *New York Times* reporter Dennis Overbye noted that this approach to the scientific sausage-making process wasn't pretty, calling it a "dissection . . . a rare example of the scientific process—sharp elbows, egos and all."[8]

It turned out that BICEP2 wasn't even the first CMB experiment to crowd-source peer review. Immediately after Penzias and Wilson's leaked results appeared on the front page of the *New York Times*, several weeks before publication of their paper, scientists confirmed it. Bell Labs was forced to issue its own press release.[9] Arno Penzias recounted the affair: "But then, the link between theory and data began to strengthen markedly once the Times article appeared. Most importantly, unexpected confirmation appeared from an unexpected direction . . . each inferring a 3 Kelvin temperature of the Cosmic Microwave Background Radiation."[10]

Three months after the press conference, in June 2014, the peer-reviewed version of the paper was published in *Physical Review Letters*. Taking the advice of two anonymous referees, we removed all trace of the dust data we took from Planck's PowerPoint slide. Its deletion, we said, was due to the unquantifiable uncertainties involved in its analysis. But we were clear: BICEP2's data were unimpeachable. It was only the interpretation that was up for debate. Planck promised to resolve the situation soon, because its newest data was set to be released in the next few months.

Planck had previously shown that the Milky Way's dust emitted microwaves with a blackbody spectrum, just like the CMB. But the dust emission had a temperature of 20 kelvin, instead of 3 kelvin. Since the total energy of a blackbody increases as the fourth power of its temperature, the Milky Way's emission was nearly two thousand times brighter than the CMB's emission.

One of Planck's channels, its frequency band at 353 GHz, was nearly insensitive to anything *besides* dust; it was a kind of sacrificial channel dedicated not to the cosmological gold we sought, but to the cosmic schmutz that might be obscuring it. We all held out hope that Planck's 353 GHz channel would be the salvation, quantifying the qualitative PowerPoint slide and allowing an unaltered conclusion. It was going to be a long, hot summer.

THE NEXT FEW MONTHS found the whole BICEP2 team on pins and needles fretting over when and what Planck would reveal. It was exhilarating. It was terrifying, too. By September, I needed to get away. As luck would have it, I was invited to give a lecture at the Italian National Institute for Astrophysics in Arcetri, Galileo's final home. If anyone could sympathize with the perils and pitfalls of the refracting *perspiculum*, it was the first person who ever pointed one heavenward.

The National Institute has the keys to Galileo's home, the Villa Il Gioiello, in the sublime rolling hills of Tuscany. For me, the thrill of eating a meal in Galileo's actual dining room was only surpassed by my view through the very same portals through which the maestro himself gazed heavenward centuries ago. (Striking my head several times on the low arched doors reminded me that Galileo's lofty stature was confined to physics.) Strolling in the villa's olive grove and grape vineyard, which literally trace their roots back to Galileo's time, was surreal. If one needs to be imprisoned, Villa Il Gioiello isn't a bad place to do one's time.

At the end of my trip, the Planck 353 GHz paper appeared. With it came the beginning of the end of the BICEP2 team's inflation ela-

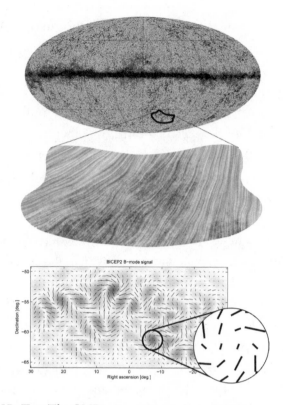

FIGURE 57. *Top*: The CMB temperature anisotropy pattern as observed by Planck, with the area BICEP2 observed (the Southern Hole) outlined in black. *Middle*: The Southern Hole as seen by ESA's Planck satellite at microwave and sub-millimeter wavelengths. The shading represents the thermal emission from interstellar dust. The "streamlines" indicate the orientation of the Milky Way galaxy's magnetic field, measured using polarized light emitted by the dust. *Bottom*: The total B-mode polarization measured by BICEP2. The total B-mode polarization is the sum of gravitational lensing B-modes and polarized dust B-modes; no evidence for primordial B-modes is present in the final, dust-corrected BICEP2 data. (© SHAFFER GRUBB)

tion. Although the Planck team was careful to release no data for the Southern Hole, the field where BICEP2 observed—perhaps out of fear we would digitize it—they made a blunt assessment of the potential amount of dust polarization contamination in the Southern Hole, saying it was of "the same magnitude as reported by

BICEP2." This meant dust was as likely a culprit for our B-modes as were inflationary gravitational waves.

Later, the Planck team produced an image of the Milky Way's dust polarization, finally including our patch of sky, the Southern Hole. It was mesmerizing; large swaths of sky festooned with azure streamers, whorls of ocher, and swaths of amber garland. Dust was showing off in all its Van Gogh vainglory (fig. 57; plate 7, top). "Visible certainty," Galileo likely would opine, as he had with his Pleiades hypothesis. But this time he'd be devastatingly right.

It was over. Eden had sunk to grief. Our Nobel gold couldn't stay. BICEP2 turned out to be a very precise dust detector. There was some schadenfreude in the community, which abounded with gleeful reports of BICEP2's demise: "Cosmic smash-up: BICEP2's big bang discovery getting dusted by new satellite data"; "BICEP2 bites the dust"; "Evidence for Cosmic Inflation Theory Bites the (Space) Dust"; "Dust thou art, BICEP2."[11] We went from being the muscle behind curls to the punchline for awful puns, with no respite in sight.

Myself, I felt both embarrassment and guilt. Although I had voiced my concerns about dust, eventually I gave in. I should have stood my ground. But, like so many of us on the team, I saw what I wanted to see, committing cosmology's cardinal sin: confirmation bias. In the end, I was Feynman's fool, and that is a role I vowed never to play again.

AFTERSHOCKS OF BICEP2

After the Planck 353 GHz paper, some scientists argued that BICEP2 had had dire effects on the public's perception of science. It wasn't because BICEP2's error meant the cyclic or bouncing models were right and inflation was wrong—it most certainly did not mean that. The jury on both inflation and the cyclic models is still out.

Rather, the post-Planck postmortem set off reverberations over the public's perception of science. Speaking of the massive media attention prior to peer review, followed by the eventual retraction of our most important claim, Dartmouth astronomer Marcelo Gleiser said the BICEP2 experience "harms science because it's an attack on its integrity. . . . It gives ammunition for people to say, hey look those scientists don't know what they're talking about in cosmology and they don't know what they're talking about in general."[12]

I disagree on two counts. What Gleiser didn't know, what no one who wasn't on the team knew, was the uprightness exemplified by the BICEP team. Jamie Bock could have used his privileged position within Planck to see their unpublished data ahead of time. But he didn't. Perhaps that's why he was so upset about me working on BICEP and POLARBEAR; the potential for conflicts of interest was too great. And in the end, the public *did* get a science lesson. Science is messy because nature is, too. Scientists are not infallible; we are human. We didn't blunder: we didn't leave the lens cap on, accidentally unplug a fiber optic link, or use dirty test tubes. Unlike Joe Weber's gravitational wave detector claims, or faster-than-light neutrinos, BICEP2's results were reproducible. In fact, we have more confidence than ever in BICEP2's actual B-mode detection significance—an amazing accomplishment, to be sure, though clearly not one for the ages. What was falsified was our assessment of their source.

BICEP2 showed the public how science works: you release a result, and other scientists work to test the result. You put your cards on the table, and leave it all out there for your critics. If and when they attack, you defend until you can defend no longer and the attacks subside. Only then, when both critic and supporter collapse, exhausted, can science be said to be settled.

In late 2014, as things began to look grim for BICEP2, science journalists convened to conduct a postmortem analysis, not of BICEP2's results but of the role that publicity played in the buildup and eventual letdown.[13] Many journalists were upset that a handful of journalists had been selected by Harvard to learn of BICEP2's results ahead of the rest. Others contemplated their role in publicizing the findings without due diligence. "How much do we emphasize the tentative nature of it?"[14] asked Michael Lemonick of *Time*, one of the few tipped off ahead of time. "It's very easy to be less than absolutely hard-nosed about your own product where an exciting announcement or an exciting news story would make a greater splash than a dull one." It was refreshing that science journalists were so introspective. They even held workshops to address their role in the BICEP2 affair, considering whether they were culpable and what should be done in the future. "It is hard to imagine political and sports reporters taking the time to discuss so thoroughly what (if anything) they did wrong after one of their stories went belly-up," read an editorial in the journal *Nature*.[15]

Science journalists do a remarkable job, usually under pressure to report on esoteric subjects with great speed. That they fretted over their role in the BICEP2 affair is reassuring. I am less sanguine about the introspection of scientists who are responsible for the science itself. Physicists rarely, if ever, receive the training in professional ethics that doctors and lawyers do. Especially considering the implications for junior scientists—whose careers will be most negatively impacted by premature or incorrect findings—this training should be mandatory, as is the training many of us now receive for preventing sexual harassment. Feynman, in his oft-quoted Caltech commencement speech "Cargo Cult Science," said a scientist should bend "over backwards to show how you're maybe wrong." Admitting you might be wrong doesn't come naturally. But then again, neither does experimental cosmology; you have to learn it. To think a scientist can intuit ethical laws is as wishful as expecting that quantum mechanics can be acquired by osmosis. Learning the best ethical practices might seem to be a distraction, but, as with most difficult things, a gram of prevention is worth a kilo of cure. The time has come to incorporate best ethical practices into our curricula, starting in graduate school, if not before.

~

FOR THE BICEP2 retraction, there was neither press conference nor viral YouTube video. And while Planck, the fearful enemy
fighter on our tail, came clean about the amount of dusty B-modes
that our galaxy produced, they never did say anything about *cosmic*
B-modes produced by inflation.

It was BICEP2's vision which was clouded: a bit by fear, a bit by
greed, and mostly by bits of dust.

⁓

I LEARNED A LOT from Jesus during my brief stint as an altar
boy. His polemic against hypocrisy, the Sermon on the Mount, has
always spoken to me as a paragon of humility: "Judge not," Jesus
said, "lest ye be judged. For with the same judgment you pronounce,
you will be judged; and with the measure you use, it will be measured to you. Why do you look at the speck of dust in your brother's
eye and pay no attention to the plank in your own eye? How can you
say to your brother, 'Let me take the speck out of your eye,' when
all the time there is a plank in your own eye?" The dust was dust
and the plank was Planck. Both had impaired our vision and our
judgment.

In late 2014, Planck would end up allowing us to see farther than
we could alone. That's when the entire BICEP2 and Planck teams
co-wrote a joint paper.[16] Competitors became collaborators, beating their adversarial swords into plowshares. Eventually, I was even
able to help recruit Raphael Flauger to UC San Diego, where he's
become one of my closest colleagues. The BICEP2 controversy was
ultimately resolved through scientific methods: BICEP2 and Planck
working together to remove the precise amount of dust producing
B-modes in the Southern Hole. What remained was an exquisitely
measured signal, but one that held no evidence for gravitational
waves, only hopes for future experiments to continue the hunt.

When Andrew Lange, mentor to the whole BICEP2 team and
a key leader of the U.S. contingent of the Planck team, shared the
prestigious 2006 Balzan Prize for Observational Astronomy and

Astrophysics from the Accademia Nazionale dei Lincei, an august Italian scientific academy, Andrew said, "Galileo joined the Accademia almost four hundred years ago. Were he here with us today, I think that he would be astounded and pleased to see the results of the program that he began when he turned a telescope to the heavens for the first time. He is quoted as saying, 'Measure what is measurable, and make measurable what is not so.' This sentiment perfectly describes our modern quest to see—and to measure—what previously could not be seen, and to extend our measurements to the furthest reaches of the universe, and to the very beginning of time."[17] I know that if Andrew were alive today, he, like Galileo, would be pleased. BICEP2 and Planck had worked together and made the cosmos measurable, in all its gritty, grimy glory.

At the conclusion of the published BICEP2 paper, we hedged our bets. Just in case the signal we saw was dust, not primordial gravitational waves, we wrote, "if these B modes represent evidence of a high-dust foreground, it reveals the scale of the challenges that lie ahead."[18] With Planck's help, after the dust settled on BICEP2, the scale of the challenges became clear: they were enormous.

The pressure was on, which was just fine by me. After all, pressure is what turns dust into diamonds.

Chapter 15

POETRY FOR PHYSICISTS

"New ways of seeing can disclose new things: the radio telescope revealed quasars and pulsars, and the scanning electron microscope showed the whiskers of the dust mite. But turn the question around: Do new things make for new ways of seeing?"

WILLIAM LEAST HEAT-MOON,
BLUE HIGHWAYS: A JOURNEY INTO AMERICA

THERE IS A LEGEND OF A MYTHOLOGICAL VESSEL, known as the Ship of Theseus, which was built and rebuilt over the centuries as the wood rotted and broke. Planks, nails, and sails were replaced one by one. Even the mast was replaced time and again. At some point, its crew wondered, when did it stop being the original, and when did it become something else? Like this legendary ship, most of the cells in our body replenish themselves—some every few days, some every decade—yet you still feel like "you." So too does our galaxy give birth to stars, which then mature and eventually die, some in spectacular supernova explosions, creating the nebular material that fuels the next generation; not on timescales of decades, but over billions of years. Is it not the same galaxy?

Even more poetically, some of the iron in our blood came from our galaxy's supernovae long ago. We are stardust reincarnate. The iron flowing through our veins first coursed through the cosmos.

RIVERS OF LIFE

For the Incas living in what is now northern South America, the Milky Way, which they called Mayu, was a life-giving river filled with two different types of constellations—luminous and dark (fig. 58). The former were made of sparkling stars but, despite their scintillating appearance, the Incas considered them to be inanimate, dead, sterile. In contrast, the dark constellations were the stuff of life itself, a silent movie with a rich and evocative foundation myth. While today, astronomers recognize these dark, absent asterisms as the result of absorption of light by particles of interstellar dust, the ancients saw them as vibrant shadows of thirsty animals: the Toad, the Llama, and the Serpent. These mythical beasts obscured the ethereal glow of Mayu as they drank from the waters of the celestial river.[1] Life playing out in dusty rivers; celestial poetry in motion.

Centuries after the Incas, the Voyager 1 space probe turned its cameras back toward the inner solar system on Valentine's Day, 1990. The probe's cameras captured the Earth bathed in starlight from our Sun (plate 7, bottom). In his book *The Pale Blue Dot*, Carl Sagan captured the insignificance of our planet, calling it a "mote of dust suspended on a sunbeam."[2]

What we've learned about other planets since Sagan's time has only magnified our insignificance. The Kepler space telescope has shown astronomers that there are at least as many planets in our galaxy as stars: hundreds of billions. Stars like TRAPPIST-1 even harbor multiple planets in the so-called habitable zone, sometimes known as the "Goldilocks zone": not too hot and not too cold for liquid water to exist.[3]

But what if the concept of a Goldilocks zone is just the latest chapter in astronomy's anti-Copernican fairytale? Who knows what kind of life we will encounter as we venture out into the Milky Way. What if all life doesn't require the same exact raw materials we do? What if carbon-based life isn't the only life form that exists? That

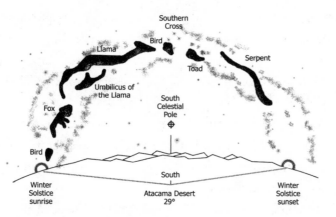

would be a natural candidate for the sixth Great Debate, one of the many more delightful debates to come.

"The seeker after truth should be humbler than the dust. The world crushes the dust under its feet, but the seeker after truth should so humble himself that even the dust could crush him. Only then, and not till then, will he have a glimpse of truth."

MAHATMA GANDHI, *THE STORY OF MY EXPERIMENTS WITH TRUTH*

BEFORE BEING COVERED in Nobel gold, Adam Riess humbled himself. He took a deep dive into dust, that often overlooked, most mercurial material that hardly would seem to be the substance of glory. In a tour-de-force 1996 paper titled "Is the Dust Obscuring Supernovae in Distant Galaxies the Same as Dust in the Milky Way?" Riess, his thesis advisor Robert Kirshner, and their collaborator William Press wrote, "Understanding the absorbing properties of dust in distant galaxies is of great importance to the measurement of cosmological parameters. . . . For the task of measuring *global deceleration* from supernovae, understanding the properties of dust in the young universe provides a formidable challenge."[4] (Emphasis mine.)

This quote from the future laureate shows clearly that astronomers in the late 1990s were biased *against* an accelerating universe. Thus, it was no surprise that they looked to confirm that the universe was slowing down. It was not only confirmation bias that initially led them astray, but also another form of bias: authority bias. In this case, it took the form of none other than Albert Einstein. Einstein dispensed with his 1917 notion of the cosmological constant, the anti-gravity he had once proposed to prevent the universe from contracting, after seeing Hubble's data on the recessional velocity of galaxies. An apocryphal story has it that Einstein called the cosmological constant his "biggest blunder."[5] For the next eighty years, people forgot about it. For good reason: why waste time looking for a blunder?

So, in the late 1990s, when at first the Type Ia supernovae appeared dimmer than expected, astronomers like Riess thought this unexpected result was due to diminution by dust, not to the more exotic explanation that ultimately proved correct: dark energy, which more than merely prevents the universe's collapse, as Einstein thought; it causes the expansion to accelerate. By first setting their sights on dust (measuring it, subtracting it from the outset), the supernova astronomers were drawn to the most unexpected conclusion imaginable: the universe *is* filled with the anti-gravitational, repulsive energy that Einstein posited and then disowned. If Einstein did indeed deem the cosmological constant his "biggest blunder," that turned out to be his blunder—not the notion of the mysterious repulsion itself.

What cosmic kismet causes the same substance—dust—to help some astronomers win a Nobel Prize while causing others to lose it? The cause, in part, lies in our biases. Whereas the supernova astronomers initially went looking for dust and found dark energy, we CMB cosmologists went looking for inflation, the most far-out explanation of them all, and found dust instead. Of course, BICEP's astronomers weren't the first to be duped by dust. But we may just be the last.

HEIRS TO CREATION

It's taken centuries for dust to get the respect it deserves. Dust shapes whole galaxies and forms entire planets; it is the bedrock of life. But dust has a dark side as well. It dims distant luminaries, fooling us into thinking the cosmos is somehow centered on us. This creator of worlds and destroyer of egos (and Nobel Prizes) remains inscrutable. The size, shape, composition, and origin of dust grains remain mysterious. The mechanism by which dust gets magnetized is elusive, and little is known about the Milky Way's magnetic field, which orients the grains, polarizing the CMB.[6]

We can't live without dust; that much is clear. The question is, can cosmologists live with it? For a mote-dwelling cosmologist to see back through the universe, to perhaps glimpse the Big Bang, means peering through all this cosmic schmutz.

One ambitious proposal to do just that calls for a novel balloon-borne telescope called "BFORE: The B-mode Foreground Experiment," led by Andrew Lange's former graduate student Phil Mauskopf of Arizona State University.[7] BFORE will specifically hunt for dust, as the supernova hunters did so successfully before they made their stunning Nobel Prize–worthy discovery. In doing so, cosmologists working on BFORE hope to unveil inflationary polarization signals that may be buried beneath the dust. Innovative projects like these do a great service to cosmology, lifting the veil on the cosmos's most mercurial actor.

"I will make your offspring like the dust of the earth."

GENESIS 13:16

WHILE IT WAS disappointing to have to retract BICEP2's claim, I can honestly say that I was not depressed afterward. The game was still on. There was one positive outcome that could never be retracted after BICEP2: humanity longed to hear both sides of the fifth Great Debate. Do we live in a single cosmos that cyclically repeats in time? Or do we live in a profligate multiverse, one with a possibly infinite number of "pocket universes"? The world can hear both sides of the Debate again, and it will be just as thrilling as the first time.

For those future debates, the stakes could not be higher. Inflation's biggest critic, Paul Steinhardt, implored teams of experimental cosmologists to go on, but with much higher standards: "This time," he wrote in *Nature*, "the teams can be assured that the world will be paying close attention. This time, acceptance will require measurements over a range of frequencies to discriminate from foreground effects, as well as tests to rule out other sources of confusion. And this time, the announcements should be made after submission to journals and vetting by expert referees. If there must be a press conference, hopefully the scientific community and the media will demand that it is accompanied by a complete set of documents, including details of the systematic analysis and sufficient data to enable objective verification."[8] Steinhardt was advising us to be fastidious, both experimentally and ethically, as we embark again on the hunt for history's oldest fossils. To Steinhardt's admonitions I add one more: agnosticism toward the Nobel Prize.

"What is defeat? Nothing but education; nothing but the first steps to something better."

WENDELL PHILLIPS

EVEN THOUGH BICEP2'S initial claim withered, it planted the seeds for the field's renewal. A half dozen or so CMB polarization experiments now had a second wind in the search for those primordial signals.

In December 2014, as the controversy settled down, Jim Simons called me for the second time that year. Fortunately for me, December's call was more relaxed than the call I got during the March BICEP2 press conference. "I've been talking to folks at Princeton. I know you're competing against them with the POLARBEAR project you and Adrian Lee are leading," Jim began. (The competition Jim was talking about was the Atacama Cosmology Telescope (ACT), run by Lyman Page, Suzanne Staggs, and David Spergel at Princeton and Mark Devlin at the University of Pennsylvania.)

"But I've been thinking," Jim continued, "why don't you guys join forces instead of competing?"

Jim asked a great question: why not get along? Imagine what we could accomplish together. He was used to asking good questions and putting his money where his mouth is to answer them. Just like Alfred Nobel, the Simons Foundation, which he and Marilyn Simons started in 1994, rewards "basic—or discovery-driven—scientific research, undertaken in pursuit of understanding the phenomena of our world."

Before he became a Wall Street power broker, Simons was a math professor.[9] In late 2014, the financial titan was suggesting a merger. The consolidation would not be of the corporate kind but of the cosmic kind, between academics who were used to competing against one another. Teams of our size had never before merged. Instead, a sort of East Coast/West Coast competition, not unlike the hip-hop rivalry of the 1990s, pervaded the community of CMB experimentalists.

The alliance Jim suggested would be a major paradigm shift. There would be birth pangs, for sure. But if any organization could catalyze it, it was the Simons Foundation, which has nucleated collaboration across diverse disciplines, ranging from genomics to quantum computing.[10]

He had a good point: having two groups a stone's throw from each other in Chile was wasteful. There was much work that was duplicative; collaboration would be a far more efficient use of the Simons Foundation's resources than two groups—however highly capable—going it alone. Collaboration is the key catalyst for many successful Simons Foundation–supported projects. (Unknowingly, the Simons Foundation had identified one of the Nobel Prize's biggest obstacles to mankind-benefiting research: the historic focus on isolated researchers, the limit of three laureates per year.)

In late 2014, after BICEP2's dusty denouement, Jim was inviting me to think big. It was obvious I couldn't do it alone. But I didn't have time to play tennis with all the principal players. So instead, I invited Mark Devlin of the University of Pennsylvania and Princeton's Suzanne Staggs to UC San Diego. Ostensibly, my invitation was just for these eminent astronomers to give a talk. But, once they were on campus, I started testing the waters, asking their opinion of what the next-generation CMB experiment in Chile would look like. How many telescopes, what frequencies would you use, what kind of detectors would be best?

After the ACT experimentalists visited, I roped in my closest collaborator, Adrian Lee of UC Berkeley. Adrian is a world-renowned experimental cosmologist and the driving force behind the innovative superconducting bolometers used in the field by POLARBEAR and the Simons Array, I had to get Adrian's buy-in if we were to move forward. And buy in he did.

13,820,000,000 + 50 YEARS

In June 2015, Princeton University held a three-day event to celebrate the fiftieth anniversary of Penzias and Wilson's discovery of the CMB. Jim Simons invited me to meet him for lunch during the celebration. He wanted to hear if I had made any progress toward his idea for a CMB "supergroup." He began to eat; I began furiously sketching on a napkin the ideas Devlin, Lee, Staggs, and I had been pondering.

For the next-generation observatory, we'd need an instrument more than an order of magnitude greater than BICEP2. While BICEP2's 512 polarization detectors were exquisitely sensitive, they were all at a single frequency. We would need detectors spread across a wide range of frequencies. That amounts to something like tens of thousands of detectors. Low frequencies would mitigate the Milky Way's synchrotron radiation, and high frequencies would measure, and allow us to remove, dust. This new super-experiment would also address other fascinating problems in cosmology: we could measure the masses of neutrinos, explore how giant clusters of thousands of galaxies form, and probe dark energy.

Finally, I told Jim, the joint project would be a prototype for an even more ambitious CMB experiment, which was then in its infancy. A proposal was brewing for an experiment that will one day, with some luck, be funded by the U.S. Department of Energy and the National Science Foundation. This future project is the stuff of cosmology dreams. CMB researchers call it the Stage Four CMB Experiment, a name only a government agency could love.[11] It could have as many as a half million detectors, operating on multiple telescopes in two locations: Chile and the South Pole.

I told Jim that to jump more than three orders of magnitude—from the number of detectors BICEP2 used to Stage Four levels all at once—was improbable. The experiment I was sketching on my napkin would bridge the gap between two generations of experiments and return fantastic science. But it wouldn't be cheap.

Then Jim opined that we would need more than just brains and cash; we would need to run the project in a much more disciplined fashion than was usual. I agreed, saying we'd need a professional project manager, something uncommon in CMB experiments. "How about your oldest kid?" he joked. "He's busy with preschool," I replied, between a smirk and a laugh.

We meandered back from lunch, talking science and reminiscing for hours beside an ivy-smothered library on Princeton's campus. I remained carefully out of earshot of potential competitors. It was a bit surreal; only a stone's throw from where Princeton's Dicke and company had been scooped out of their Nobel Prize by Penzias and Wilson, I was busily scribbling ideas that might lead to truly beneficial findings someday, even if they were meticulously orchestrated instead of serendipitously discovered.

As the day drew to a close, Jim asked me one more question: "How many years before it's in the textbooks, if you're right?" I said maybe ten, but less than twenty. He said, "Let's make it closer to ten!"

~

THE REST OF that year, the leaders of the Simons Array and ACT teams spoke weekly, hammering out the details of what would become the next chapter in experimental cosmology. As usual, another critical task was to come up with a name. Channeling Galileo's obsequiousness, we christened it the Simons Observatory, assuring our patrons that revolutionary astronomical discoveries would come later.

Next, we thought long and hard about the technical details. We worked on a design, budget, and schedule to present to the Simons Foundation. There would be multiple telescopes: a mixture of large telescopes, like the 5 m (16-foot) diameter ACT telescope, and smaller ones, similar in size to BICEP's 30 cm (1-foot) diameter spyglass. We'd keep the same Atacama Desert site, because it is one of the world's premier astronomical observing locations. The Observatory would need a broad spectrum of frequency channels in order to measure and remove dust.

We wrote a set of founding documents detailing the legislative structure of the project, full of checks and balances, with oversight both within and from outside the project. The External Advisory Committee is a group of eminent scientists who are charged by the Simons Observatory to give us unbiased advice. Their technical guidance alone is critical. But more than that, they provide impartial oversight that we hope will avoid some of the painful lessons learned from the BICEP2 experience.

The Simons Observatory's founding documents include a charter, a leadership board, and rules to run the collaboration. I became the director of the Simons Observatory, with overall stewardship of the Foundation's resources. The project's day-to-day leader is its "spokesperson," a position that rotates every two years. Our first spokesperson is Penn's Mark Devlin, who will be followed by Berkeley's Adrian Lee, and then Princeton's Suzanne Staggs. Being spokesperson is every bit as challenging as being the CEO of a fairly large company; I have no doubt, though, that Staggs, Devlin, and Lee are so enormously talented that each could run a Fortune 500 company, were they so inclined.

The marriage of two experiments would immediately bring together nearly two hundred people, almost four times more than worked on BICEP2. We submitted the proposal to the Simons Foundation in late 2015, and in early 2016, after I gave a talk to the Simons Society of Fellows in Manhattan, Andy Millis, the Simons Foundation associate director for physics, broke the good news: "We have decided to fund the Simons Observatory."[12] I was ecstatic. The next order of business was, of all things, a press release.

The parallels between the Simons Observatory's press release and BICEP2's media frenzy tickled me. News outlets around the world carried the press release, as they had for BICEP2.[13] And for good reason—the stakes are as high as ever. This time, I was at the heart of it. But what touched me most was to be working with cosmology's best and brightest, young and old, many of whom used to be competitors. Instead of competing, we were all working together toward a common goal: learning how it all began and, perhaps, how it all will go.

POETIC JUSTICE

Dust has clouded all five of the Great Debates. And it will continue to do so if we continue to make it the villain each and every round. Sitting here on this planet, this giant dust grain, it's only natural that the most evolved form of dust would overemphasize its importance. It would be foolish to declare that we've witnessed the final Great Debate; doing so would perpetuate the cosmic hubris that has so often

FIGURE 59. The Simons Observatory collaboration. (PRAWEEN SIRITANASAK)

been our downfall. Dust has been the astronomer's most reliable fellow traveler. So, instead of seeing dust as some sort of cosmic virus, as merely our galaxy's "dirty windshield" obscuring the view of the road ahead, we should embrace the dust. For it could be true paydirt.

In 1960, John Updike wrote a paean to neutrinos called "Cosmic Gall."[14] Updike's poem was featured on the Nobel website in 1995, in commemoration of the discovery of neutrinos and the subsequent Nobel Prize won by Reines and Cowan. Updike compares these elementary particles to dust maids. But at least dust maids interact; neutrinos are the snobs of the high-energy physics society, snubbing sensors, seemingly too elite to interact.

In contrast, dust is the hardscrabble foundation of the world. Dust is as gritty as the common man. Dust is fundamental, if not elemental. While cosmologists delight in constructing exotic models to explain the universe's dark matter and dark energy, entities which are noncommunicative in the extreme, neither emitting or absorbing light, they diss dust, the most ubiquitous, sociable, cosmopolitan form of matter there is.

This got me thinking, dust deserves its own homage! So, to rehabilitate dust's image, to extoll dust's many virtues, to repay debts due to dust, I hereby offer my own poem in praise of this unsung hero of the cosmos.

Conscious Starstuff

Dust grains are mysterious things.
Attached to your toddler they surely bring
Grime and grunge in rooms unkempt
Signs of love and life well spent.

From a star's nursery dust arose,
Only to be belched out in its death throes.
Through space we sail on a cosmic mote,
Trying to read the first prologue ever wrote.

Tracked by shoes into the cellar,
and riding upon winds interstellar.
Dust allows a fleeting existence.
But it pays to scrub it with persistence.

Dust covers playgrounds, filled with laughter,
and accompanies the sarcophagus to the hereafter.
We were warned long ago from the Mount,
"For thine own dust thou shall account."

Supernova slag flows through our veins;
dust causes worlds to wax and wane.
Iron filings, pyroxene whiskers;
Silicate shavings squelch Nobel whispers

It can't be emphasized enough:
The dust is us, conscious starstuff.

Chapter 16

RESTORING ALFRED'S VISION

"Finally, it is my express wish that following my death my veins shall be opened, and when this has been done and competent Doctors have confirmed clear signs of death, my remains shall be cremated in a so-called crematorium."

THE FINAL SENTENCE OF THE WILL OF ALFRED NOBEL

IMAGINE THAT, AFTER MUCH "MATURE DELIBeration," you methodically delineated who should get what, how much, when. You left money to your family, to your friends, even to your servants. How might you feel if upon reaching the Pearly Gates you found out that your specific and explicit intentions had been ignored, or worse yet, tampered with? You'd likely be outraged. But there's no recourse beyond the grave. What could you do?

We began our story with the specter of death. The deaths of three of the original seven Nobel children gave rise to the strange awards ritual repeated annually on the date of Alfred Nobel's demise. His final testament closed with a macabre request that resulted from his taphophobia, the fear of being buried alive. It was a paralyzing condition, one he'd inherited from his father.[1] Opening his veins might help determine the cause of his demise, but more importantly, it would prevent a premature burial. While Alfred wasn't buried alive, he was immortalized in a way he scarcely could have imagined.

The Nobel Prize is perhaps the most prestigious award, not just in science but of any kind. It has aimed to recognize humanity's crowning achievements in the hard sciences, free from ideological agendas or fashion. When it's "done right," it is a truly lustrous meritocratic reward system, inspiring scientists and laypeople, both young and old. Winning the Nobel Prize confers instantaneous celebrity. Sometimes winners become household names, if only for a week. To physicists, laureates achieve permanent fame. Indeed, some Nobel Prize winners achieve an almost demigod status.

But, despite his foresight, the prizes have so dramatically changed that Alfred Nobel might not recognize them were it not for his name. The Nobel Prize today is a reward mechanism that discourages collaboration, celebrates authority, and sets up a rat race where claim-staking, speed, and greed are encouraged—the distorted vision caused by what I've called the prize's three broken lenses. To these already damaged spectacles, I'd be remiss not to add one more Nobel deficiency: its myopic focus on discoveries made by male physicists.

WOMEN AND THE NOBEL PRIZE IN PHYSICS

With no progeny of his own, Nobel showed avuncular generosity to the children of his brothers. His nephews were especially enriched; they received twice the inheritance of his two nieces.[2] Despite this inequity, Nobel never specified that his prize couldn't be awarded to women. But it almost seems like he did.

There have been nearly as many Nobel laureates (207) in physics as Popes serving the Vatican (266). Yet, there have been only two more female physics laureates than female Popes. Even the award to the first woman physics laureate, Marie Curie, was made grudgingly; only after only righteous protestations by a Swedish Nobel committee member, Magnus Mittag-Leffler, and her co-laureate,

FIGURE 60.
Maria Goeppert-Mayer
at the University of Cali-
fornia San Diego. (MARIA
GOEPPERT-MAYER PAPERS,
SPECIAL COLLECTIONS AND
ARCHIVES, UC SAN DIEGO)

Pierre Curie, did she win the prize she deserved. A six-decade female-free drought ensued until UC San Diego's Maria Goeppert-Mayer became the second and to date only other female physics laureate. Even Goeppert-Mayer's award was tinged with male chauvinism; the San Diego newspaper reported her victory with the headline "San Diego Housewife Wins Nobel Prize."[3]

Currently, women make up an unfortunately small fraction, approximately 20%, of the physics faculties at major U.S. research institutions. But that fraction dwarfs the percentage of female physics laureates by a factor of twenty. All other fields—including economics, the newest comer on the Nobel circuit, with one female laureate compared to seventy-five male winners—have a higher percentage of female laureates. Physicists are now asking themselves how the lack of gender diversity is affecting the career choices of young women. And even Nobel Prize winners like Brian Schmidt are speaking up against the prize's lack of gender diversity.[4]

The dearth of female physicists affects male physicists too. A 2017 study interviewed 121 British and American physicists and found that female scientists are perceived to be more ethical than their male counterparts, stating, "The severe competition in science contributes to scientists' ethically gray conduct, such as the temptation to pressure students or do scientific research and publish results too quickly, all in an effort to help scientists get ahead of the competition."[5] The same study found that women may be less subject to the negative effects of competition than their male counterparts. Studies like these support my claim that competition is negatively affecting physics research.

While I'm proud to have mentored several outstanding female PhD students (all of whom have gone on to win the most prestigious postdoctoral fellowships in physics), I'm only playing a small role. The lack of diversity in prizewinners gives the message to a young woman deciding on her choice of profession that in physics women are not equally valued. A vicious cycle results in which women fail to enter the field, denying younger women role models; it is the anti-Matthew effect. Women disproportionately miss out on the Nobelist's noblesse-oblige phenomenon where "Scientists who as young men worked with a laureate received the award at an average age of forty-four, nine years earlier than men who had not."[6] Particularly egregiously, although Jocelyn Bell studied with a Nobel prizewinner (1974 Nobelist Antony Hewish), she never received the "laureate-begets-laureate" benefit. Indeed, many have argued that Bell should have received a share of the 1974 prize for making the discovery for which Hewish was rewarded.[7] (Once again, there is no written rule why Bell cannot be awarded her fair share of the 1974 prize *today*; there is only the arbitrary precedent of it not having been done before, for fear of condemning prior Nobel committee members.)

Unmooring the prize from Alfred's "the person" bonds would happen if the physics prize were awarded to groups. This would reduce the pressure on scientists to stake their claims at the expense of others; it would offer a shortcut up the ladder of authority, a ladder some underrepresented, and thus less powerful, groups such as women and other minorities feel has already been pulled up out of reach.

AN OCTOBER SURPRISE

Eight months after submitting my nomination for the 2016 Nobel Prize, I stayed up all night until the prizes were announced on October 4, 2016. To a physicist, the Nobel Prize announcement is like election night combined with Super Bowl Sunday. There's always a feeling of anticipation crackling around the department. I secretly hoped to share the reflected glow if my nominees won.

The prizes were announced at 2:45 a.m. Pacific time: Michael Kosterlitz, David Thouless, and Duncan Haldane won the 2016 Nobel Prize "for theoretical discoveries of topological phase transitions and topological phases of matter." They were not my nominees. But I was happy for them just the same; Kosterlitz was one of my professors at Brown, and Duncan Haldane had been a professor at UC San Diego when he did his prize-winning work.

It didn't take long for popular media to bemoan the complexity of the work. It was fun to see the Nobel committee's gymnastics as they described topological matter in terms of donuts and coffee cups and pretzels. It was a good prize, awarded to a serendipitous theoretical discovery and to the primary three scientists who discovered it. That was the good news.

I'd be lying if I said I wasn't a bit disappointed. I knew my candidates would probably need to wait at least five more years, since the field rotates from year to year across all fields of physics; it's almost unheard-of for the same branch of physics to win more than twice in any ten-year period. While, obviously, this is not part of the Nobel statutes, nor was it included for my consideration as a nominator, nevertheless it's apparent from the list of Nobel Prizes over the decades. It is yet another arbitrary stricture that shows the prize has changed from what it was intended to be—for the most beneficial discovery made in the preceding year.

My astrophysicist nominees, whom I will not name to protect the innocent, were likely ineligible to win for perhaps another five to ten years. It seemed inevitable that the LIGO scientists would win in 2017, as indeed happened.

> *" 'Do not be angry, my Lord,' Aaron answered. 'You know how prone these people are to evil. They said to me, 'Make us gods who will go before us.' . . . So I told them, 'Whoever has any gold jewelry, take it off.' Then they gave me the gold, and I threw it into the fire, and out came this calf!"*
>
> EXODUS 32:22

I ALWAYS THOUGHT the "Golden Calf" episode was the most credulity-stretching story in the Old Testament. How could the Israelites make themselves a false god just weeks after seeing the real God utterly devastate Egypt courtesy of ten supernatural plagues?

If that was not enough, surely splitting the Red Sea a few days later should have convinced them to listen to the Miracle Worker in Chief: do not make false idols. It's the Second Commandment, for Godsakes! The whole story was completely irrational. It caused me to doubt the wisdom of the Good Book. If the Bible was supposed to be factual, its description of human nature sure seemed fictional. Who could possibly worship a golden god that they themselves had forged? Well, I thought, perhaps everything the Bible had to say was hopelessly irrelevant in our modern scientific age.

Then came May 11, 2017, the day I witnessed society's brightest minds, its most secular and arguably most logical members—physicists—worshipping a golden idol. That Thursday, Duncan Haldane returned to give a lecture at UC San Diego, and he brought his Nobel Prize medallion for all to see. A throng of physicists young and old got caught up in the excitement. They fawned over the gilded medallion, pushing past one another just to get a glimpse of it. No one actually bowed down to it. But some did kiss it. Some tried to

FIGURE 61. *Left*: The author with 2016 Nobel laureate in physics F. Duncan Haldane. *Right*: The author posing with the last Nobel Prize medallion he will ever touch. (BRIAN KEATING)

sneak off with it. I'm ashamed to say I was among the worshippers, unable to resist posing for a picture with the medallion. I had been writing this book about the prize's idolatrous power to refract reality, and there I was, venerating the very golden graven image whose menacing aspects I had been wrestling with for the better part of a year.

Perhaps we're done with coveting our neighbor's asses, but idol worship is deeply rooted in human nature, and human nature hasn't changed much in the past few millennia. Looking at my brush with the actual Nobel Prize, I realized that modern man is not so far removed from the exalters of bull and Baal in centuries past. What do physicists venerate in our modern age? No Baal statue is worthy of worship nowadays; no bull either. But there is Nobel.

I began to note similarities between the secular religion of "Nobelism" and other monotheistic faiths. Nobelism's patriarch was Alfred, son of a capricious, sometimes tyrannical father. After the father abandoned the son in his youth, Alfred went from Scandinavia to Paris, where he thrived. But he was persecuted and exiled from his Promised Land. Eventually, with his health failing, he took a hero's journey, slipping back into his beloved Paris to inscribe an everlasting testament with the faith's catechism "for the benefit of mankind." Nobelism has its patron saints (the laureates), its high priesthood (the committee), two holy days—the Call of Revelation (in early October) and the (Reindeer) Feast of Coronation (December 10)—and, of course, no shortage of graven gilded icons.

And there is one last parallel between Nobelism and religion: conversion. If you've ever been approached by a Jehovah's Witness, you'll know how resilient one's choice of religion is, especially so if one is secular. Religions give people's lives structure and meaning; elevating them to a higher moral plane. Religions convey authority, a status that, once cultivated, must be maintained.

As I wrote this book, some scientists implored me to cease my apostasy, telling me that my recommendations would dilute, maybe kill, the Nobel Prize, certainly rob it of the luster that made me lust for it—a desire that, by my own admission, played a role in my creation of BICEP. But that advice missed the point of this gentle jeremiad. I don't want to crucify the prize. I want to revitalize it, albeit in a reformed way. To do so, I asked myself, "If Alfred were resurrected, what should he change in his will?"

It turned out there were many changes that were needed. Best of all, they all had precedent.

ESTATE PLANNING SOLUTIONS

If Alfred Nobel came back, he might be shocked at how far we've strayed from his dreams for a more peaceful world catalyzed by his prizes. The Nobel's first century saw peace prizes awarded to bona fide terrorists and warmongers; a prize in the economic sciences, never authorized by Alfred, and one that has become highly politicized;[8] the literature prize awarded to a popular musician; and perhaps most troubling of all to Alfred, chemistry prizes awarded to, gasp, biologists (he was a chemist, after all). What would Alfred have to say about the performance of his will's executors, the Nobel Foundation?

An estate attorney examining the most famous will in history might conclude that the foundation has done well in many regards. Certainly, it has grown the financial endowment significantly, and the prize amounts have trended steadily upward. The prize's prestige is at an all-time high. More people than ever tune in to watch the awards live on the Internet.

To answer the question more fully, we must try to understand what Alfred intended with his bequest. The motivational power of prizes has been proven so many times, the effect is almost self-evident. Nobel Prize historian Burton Feldman has argued that the timing of the inaugural Nobel Prize in the early twentieth century was particularly fortuitous, marking the simultaneous ascendancy of the media and the beginning of science's modern age, an age when science became "increasingly incomprehensible to the public, and also when the media began its own expansion and influence."[9] Alfred Nobel bequeathed a fortune to scientists, activists, physicians, and writers in order to draw attention to science, not to turn scientists into millionaires or celebrities.

Yet attorneys looking deeper might find that the prize has rarely, if ever, conformed to the actual strictures of the will. There have been many changes in the past century, but three most significantly affect the physics prize today.

The first modification came in the prize's second decade, when the number of winners in any one year was limited to three. Alfred's will indicated that only a single laureate win each prize, and at first blush it seems the Nobel Foundation was being charitable in allowing up to three scientists to share it. However, as Elisabeth Crawford shows, the original statutes the Nobel Foundation employed from 1901 to 1915 were more generous, potentially allowing groups to win (as the peace prize allows to this day), not a mere one, two, or three.[10]

The second, and surprisingly most controversial, addition came in 1968 when the Nobel Prize in Economic Sciences was first awarded.

It was later rebranded with an admittedly winsome new name: the Sveriges Riksbank Prize in Economic Sciences in Memory of Alfred Nobel. Why? Because it had become distorted into a "public relations coup for economists to improve their reputation," according to Peter Nobel, Alfred's great-grandnephew.[11] Nevertheless, the prize is still commonly referred to as the Nobel Prize in Economics and its winners are as lionized as are the winners of the original five Nobel Prizes.

The third major change came in 1974, when the Nobel committee forbade posthumous winners.

Why do some Nobel Prize rules change and others stay the same? Why do some rules apply to some prizes, but not to others? Because human beings make the prize, and we are subjective by nature. Prizewinners aren't selected by complex algorithms that weight a scientist's contributions, methodically employing sophisticated metrics to determine suitability; no such purely objective mechanism exists.

Scholars such as Robert Friedman, writing in *The Politics of Excellence*, have depicted the Nobel committee as a capricious cabal, often awarding the prize "to promote themselves and the ideas in which they believed." My colleagues and I have a bit more regard: the majority of physicists I've spoken to agree that most of the laureates deserved their prizes. It's sublime when the committee gets it right. But it's insufferable when they are biased, or adhere to arbitrary strictures, hidden behind opaque protocols, seemingly more concerned with adhering to the letter of their (not Alfred's) law than to its spirit. Never before has such a small group of people had such outsized influence on humanity's perception of science. With such power comes a tremendous ethical burden.

THE NOBEL REFORMATION

Nearly fifty years have elapsed since the last substantive changes to the statutes of the Nobel Foundation were made. Even the U.S. Constitution must be amended every now and then to reconcile the founders' intent with the realities of the modern world. A visionary like those founders, Alfred Nobel would no doubt be proud if his will, ironically, became a living document, aligned with the current state of science while continuing to influence humanity to the good.

That a prize conceived in the nineteenth century still motivates scientists in the twenty-first century is a testament to its allure, for both scientists and lay observers. Early on, the prize suffered from nationalism, outright racism, and sexism. Much has changed for the better. But if the prize is to remain germane to modern scientific enterprises, an overhaul is needed.

Many scientists are daunted by the prize's ponderous pace. Given the long odds against funding, who can blame newly minted researchers for looking glumly at the decades-long prize process? And, as we've seen, even some laureates feel their victory would not be possible in today's climate. Younger scientists—those I expected would desire the prize most—are becoming disillusioned by the lack of diversity of the winners, saying the gilded prize no longer reflects the broad panorama that characterizes modern science. Some have called for a boycott of the prize until it reflects modern science's inclusive and massively collaborative nature.[12]

Instead of boycott, I propose reform. In the spirit of Lutheranism's founder, Martin Luther, I submit (ninety fewer) theses to the Nobel committee in hopes of achieving a reformation for the Church of Alfred Nobel.

1. Add prizes in vibrant new scientific disciplines.

Doing so wouldn't reduce competition within a field per se, but it would spread the wealth around and recognize up-and-coming interdisciplinary fields. When the economics prize was established in 1968, it in no way diminished the impact of the Nobel Prize in Physics or Chemistry. Neither would adding new disciplines now, such as artificial intelligence, or quantitative biology, or even biology itself! As physicist Jim Al-Khalili says, "Why not just reward the best research, rather than pigeonholing disciplines? After all, it's not a new idea; physicists and biologists have worked together fruitfully in the past. Didn't Crick (a physicist) and Watson (a biologist) do just that?"[13] Increasing the attention to, and funding for, science, now more than ever, would offer immeasurable benefit to mankind.

2. *Award the physics prize to groups of any size.* Failing to credit most of the human capital in a large collaboration is inaccurate, demoralizing, and unjust. Reopening the Nobel Prize in Physics to groups would remove an artificial stricture the committee has bound itself with, and spur precisely the types of massive collaborations that are integral to modern science.

3. *Give retroactive awards.* Instead of pretending that previous committees were infallible, correct their mistakes. They're only human, after all, and humans are known to be subject to pernicious, non-scientific forces like authority bias, wherein past accolades beget future recognition. Likewise, future committees can correct past instances of the "Matilda effect," Margaret Rossiter's term for the phenomenon wherein men get credit for discoveries that were made by women. The history of the Nobel Prize is replete with examples of this, from Rosalind Franklin's lost credit for co-discovering DNA to Lise Meitner's snubbing after she discovered the foundations of nuclear fission. Of course, these women have long since passed away—which brings us to our next, much-needed reform.

4. *Allow posthumous laureates.* Alfred Nobel wanted to encourage speedy dissemination of beneficial discoveries, not research done in haste to meet arbitrary deadlines. Yet, while the "preceding year" clause in his will has rarely been adhered to, the Nobel Committee for Physics treats the "posthumous stipulation" they added in 1974 as sacrosanct—as inviolate as a law of physics. This strict adherence demeans the accomplishments of deserving scientists who do not survive the decades-long confirmation process, and adds yet more pressure to scientists to make life-changing discoveries quickly. Great science sometimes takes a great amount of time to accomplish and then achieve acceptance. The contributions of too many deserving physicists are not being honored due to age, misfortune, or the committee's own failure to recognize a deserving scientist. Awarding Vera Rubin the first posthumous prize would be immensely inspiring to young physicists, and specifically to women. But even if the committee is unwilling to restore posthumous eligibility, it should ensure justice for 1974 Nobel Prize Matilda effect victim Jocelyn Bell Burnell; the prize was awarded to her thesis advisor for the serendipitous discovery of pulsars which she made. Thankfully, Bell Burnell is still very much alive.[14]

5. Recognize that true discovery is serendipitous. Giving prizes primarily for unexpected discoveries, as opposed to confirmation of previous theories, will incentivize more novelty within physics, bringing with it more of the high-risk, high-reward findings that have true breakthrough potential.

WHERE THERE'S A WILL THERE'S A WAY

There will doubtless be those who say my Nobel critique is born of sour grapes. There may even be benefactors, in foundations both public and private, who will say I should have spent less time writing about the Nobel and more time doing science that could win it. Perhaps they are right. But this emancipation, courtesy of a simple realization, has allowed my future forays within experimental cosmology to be carried out with Nobel agnosticism, free from science's Stockholm syndrome, however harmless it may appear. *The science is the prize*, and it's a divine compensation indeed.

AN ETHICAL WILL

"Dr. Alfred Nobel, who became rich by finding ways to kill more people faster than ever before, died yesterday."

LUDWIG NOBEL'S OBITUARY, MISATTRIBUTED

ALFRED'S SHOCK AT READING HIS FALSE OBIT-uary may have subsided when he realized it was actually a gift, not only to him but also to future generations. It was the same present Ebenezer Scrooge received forty years earlier in Charles Dickens's 1843 *A Christmas Carol*: an encounter with death that led to rebirth and redemption. Alfred died childless, with the titular prize his only heir. The prize immortalized the Nobel name, inextricably linking it to science, literature, and peace, and making it synonymous with the titanic contributions the awards commemorate. From swords to plowshares. One man. One vision. Eternity indeed.

As my brush with the Nobel Prize concluded, I came to recognize that Alfred's will was also what, in Hebrew, is known as an "ethical will" (*tzava'at*). There are many examples of such documents, ranging from presidents to ordinary citizens to Jesus himself.[1]

An ethical will is half autobiography and half legacy. Instead of the material bequests found in an ordinary last will and testament, an ethical will passes on wisdom to one's heirs, both biological and ideological. Often, the author is also the beneficiary of his or her own ethical will—another stark contrast to an ordinary will. Writing an ethical will clarifies the author's life's meaning. So, I began to contemplate my own ethical will. What wisdom would I leave behind? What ethical goals might I pass on? As an academic, I began with another close reading of a primary source, in this case the original ethical will, found in the Talmudic tractate known as "The Ethics of the Fathers." It turned out it was also a manifesto for making scholars.

> Moses received the Teaching from Sinai and gave it over to Joshua. Joshua gave it over to the Elders, the Elders to the Prophets, and the Prophets gave it over to the Men of the Great Assembly. They would always say these three things: Be cautious in judgment. Establish many pupils. And make a safety fence around the Teaching.[2]

Remarkable as this ancient "family tree" is, it only goes back five generations; my academic version (fig. 62) traces back seventeen generations, when my earliest doctoral ancestor acquired his first pupil in the early 1600s, just as Galileo began scanning the skies over Padua.

The Sage's advice to be cautious when judging is also practical, lest you be judged recklessly. So too is the advice that teachers should have many pupils. To establish something, however, suggests we are builders, in a sense. What do we build? It reminded me of Alex Polnarev's first science lesson to me: the etymology of the word "scientist" itself. In Russian, he said, scientist means "one who was taught"—a weighty responsibility indeed. But we intellectual contractors must teach more than mere facts: we must be worthy of the trust our students, and theirs to come, put in us. In short, to establish them, we must earn them.

But what does the Talmud mean by making a fence around our teaching? Eventually, it dawned on me. Fences are meant to protect; some do so by keeping out, yet the most precious fences guard what we want kept *in*. A last will and testament is a fence to preserve and protect material wealth. But do any of us truly own our property? Aren't we just the caretakers of it while we are alive? We have neither say nor recourse as to how it is actually treated, saved, or

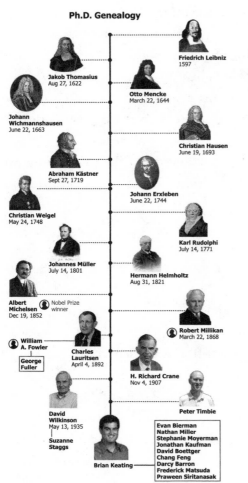

Ph.D. Genealogy

Jakob Thomasius
Aug 27, 1622

Friedrich Leibniz
1597

Otto Mencke
March 22, 1644

Johann
Wichmannshausen
June 22, 1663

Christian Hausen
June 19, 1693

Abraham Kästner
Sept 27, 1719

Johann Erxleben
June 22, 1744

Christian Weigel
May 24, 1748

Karl Rudolphi
July 14, 1771

Johannes Müller
July 14, 1801

Hermann Helmholtz
Aug 31, 1821

Albert
Michelsen
Dec 19, 1852

Nobel Prize
winner

Robert Millikan
March 22, 1868

William
A. Fowler

Charles
Lauritsen
April 4, 1892

George
Fuller

H. Richard Crane
Nov 4, 1907

David
Wilkinson
May 13, 1935

Peter Timbie

Suzanne
Staggs

Brian Keating

Evan Bierman
Nathan Miller
Stephanie Moyerman
Jonathan Kaufman
David Boettger
Chang Feng
Darcy Barron
Frederick Matsuda
Praween Siritanasak

FIGURE 62. The author's academic genealogy. (© Shaffer Grubb)

squandered after we pass away. So, too, our science belongs not to us alone but to the universe; we scientists simply safeguard it for the next generation. If we can protect what's worth keeping in, the chain of transmission can truly stretch into the far unfurling future.

A TIP FROM ALBERT EINSTEIN

Legend has it that, when Albert Einstein learned he'd won the 1922 Nobel Prize in Physics, he was in Japan and a bit strapped for yen.

After a meal, he was unable to leave a tip. In lieu of cash, he wrote a note to the waiter telling him that if he was lucky, it would be worth more than any change from his pockets. The note said, "A calm and modest life brings more happiness than the pursuit of success combined with constant restlessness." In 2017, the note sold at auction for $1,560,000, more than the current cash value of the Nobel.[3]

In this note, Einstein, the man considered the father of modern physics, gave an abbreviated ethical will, one which I realized was the perfect legacy to leave to my students.

~

TO MY SCIENTIFIC children, after mature deliberation, I bequeath the wisdom that you should pursue your scientific dreams without ambition for awards and accolades. Prizes are nice, but they can lead to the endless pursuit of success which Einstein decried— an activity which, I can tell you from experience, can become all-consuming and self-destructive.

Instead, be humble. We scientists feel pressure to know everything, to have all the answers. But we should not consign ourselves to being walking Wikipedias. There is much we don't know, and that's the real reward of being a scientist. Success isn't about knowing the answers. It's about having the questions.

MOURNING, MANHOOD, AND MELANCHOLIA

Sigmund Freud said a man "doesn't become a man till his father dies." By his definition, I became a man on June 10, 2006. It was a Saturday, the Sabbath, Shabbat. I spent it in a hospital rather than a temple, contorted into the shape of a W between two chairs in my father's room. It wasn't the way Shabbat was meant to be observed.

Shabbat had become my only source of sanity since learning of my father's awful diagnosis six months earlier at the South Pole. It was a mandate to unplug and enjoy spiritual and physical delights. Rabbi Abraham Joshua Heschel said of it, "The meaning of the Sabbath is to celebrate time rather than space. It is a day on which we are called upon to share in what is eternal in time, to turn from the results of creation to the mystery of creation; from the world of creation to the creation of the world."

If ever there was a religious duty for a cosmologist to practice, this was it. Obviously, abstinence from technology is not practical for everyone. Nor is observing the Fourth Commandment psychologically palatable to many, including many of my divinely-inspired secular colleagues.

But for me, that Shabbat began where the previous night had ended. I was with my brother Kevin in Cedars Sinai hospital, where our father was battling stage four cancer. He'd just endured the latest in a string of unsuccessful surgeries. To ease the agony, his doctors had induced a coma. He was unresponsive, but as unscientific as it sounds, it seemed he sensed my presence.

Being with him, I thought of how some people could observe the commandment to refrain from working on Saturdays by being lazy or turning to drugs, fulfilling the Sabbath technically but not spiritually. So many people sleepwalk through life itself, avoiding attachment to avert life's inevitable pain.

But then I laughed. "Dad," I said, "this is the most religiously you've ever taken the Fourth Commandment." It was the first of many jokes he'd never respond to. The funniest, most brilliant, most brutally complex man I'd ever known was rendered silent by an invisible foe. I kept talking to him for hours as if he could hear me. For all I knew, he could. As the day wore on, his breaths became shallower, more ephemeral. It was obvious he was slipping away.

He'd once told me he wasn't in the room when I entered the world thirty-four years earlier. On that day in June 2006, I was the same age he was when I was born. We were so similar, both entranced by the magic of science. Both of us had more than a little distaste for authority. We'd both been abandoned by our fathers and not spoken to them for years, the opposite prescription of "The Ethics of the Fathers." But now I'd resolved to break that particular chain. What separated us was my conviction that I *had* to be by his side simply because I was commanded to be.

I didn't have to love him (though I did). I didn't have to respect all his life choices (I didn't). But I did have to honor him. This was my way. Whatever past injustices he'd committed, however many missed graduations, driving lessons never taught, shoulders not cried upon, none of that mattered. Grievances receded at the speed of light. If I couldn't forget, forgiveness would have to do. It was all I had left to give.

Kevin and I took turns by his side. Around midnight, I left the room and cried alone. Emotional highs and lows, from the adventure of a lifetime at the South Pole to the most agonizing moments of my life, all wedged into mere months; all four seasons of life pursuing me as relentlessly as time itself.

A few hours later, I headed to my mom's house in West L.A. to get a few hours of sleep, not knowing how much longer he might go on in that state of suspended animation. The weight of it all finally hit me. Just as there was no instruction manual for raising children or being a mentor, there was no good guide for handling the death of a parent.

As I got to my mother's house, Kevin called. Dad's breathing was becoming more and more labored. I headed out again, driving Dad's old silver Porsche along Santa Monica Boulevard back to the hospital. As I carved through the empty streets I couldn't help smiling as I recalled him saying he'd gotten the Porsche years ago as part of his "end-of-life crisis."

I raced eastward, slingshotting around corners, magnetically drawn to his side. I needed to get there before it was too late. But I missed my chance. He was gone. He'd missed my entrance to the world and I'd missed his exit. For all the years I wondered who he was or where he was or why he'd done the things he'd done, in the end no questions were necessary; no answers would have sufficed.

I was thirty-four years old, fatherless again, just as I had been nearly three decades earlier when he left the first time. This time, though, it wasn't his choice to leave me. This time, I wasn't relieved to see him go. This time, I wasn't angry with him. This time, it hurt. Brutally.

My father died before his time, without writing an ethical will. But, thankfully, I had spent the most precious moments of irretrievable time by my father's side. In honoring him, it was I who received the greatest gift imaginable. Who knows whether or not I'd added days to my life, as, perhaps, the Fifth Commandment promised. But I had added life to my days, and that was reward enough for me.

> *"Life can only be understood backwards;*
> *but it must be lived forwards."*
>
> Søren Kierkegaard

We astronomers are science's ultimate scavengers, making do with whatever the universe gives us: a few photons here, a cosmic ray or two there. Occasionally a meteorite or a gravitational wave gets thrown in as a sort of cosmic seasoning, whetting our appetites for more. But there is one advantage we astronomers have over all other branches of science: we have our telescopes—the ultimate time machines. As I grew from an amateur to a professional astronomer, I marveled at how my curiosity magnified . . . as did the size of my telescopes. The stakes grew, too.

Though what we discovered was more mirage than multiverse, BICEP2 still revealed more than I had ever thought a telescope could. I still marvel at what I glimpsed through it. I went to the ends of the Earth. I eulogized my father and the man who was my scientific father. I was hired and fired and hired once again. I met the love of my life and welcomed three precious souls to the world. I nearly grasped the gilded medallion, and felt it slip away. BICEP2 transformed me; I emerged a more passionate scavenger than ever before.

The cosmos's prologue is still there to be read, as long as we can get past its dust jacket.

~

ONE NIGHT, NOT too long ago, I took my children outside with a small telescope not unlike the one I first used as a thirteen-year-old. Through it, we gazed at the stars, the Moon, and Jupiter—the same sights that had ignited my love affair with the night sky. Entranced by the heavens, we hardly noticed time's passage. But soon it was time for bed. As I tucked them in, I heard them debating the merits of the stars, their colors and intensities, which one was the best, and why. My only response was that oldest song of astronomical awe: "Twinkle, twinkle, little star, how I wonder what you are. . . ."

After they fell asleep, I took the telescope outside one more time, alone. Looking up, I was a kid again—just a boy at home under a canopy of silent stars. My telescope was my time machine once more, connecting my mind's eye to my past self. Suddenly, I was overwhelmed with emotion. I had found a way to lengthen my days. I'd had the time of my life. My childhood angst was gone now. In its place, reality was reborn: life's mysterious meaning refracted and clarified by lenses of awe and wonder. It had all happened so fast.

> The universe unfurled around me.
>   Time wasn't on my schedule.
>   But it finally made sense.
>   I just had to view it in reverse.

ACKNOWLEDGMENTS

This book was inspired by my colleagues young and old. A particularly large debt of gratitude is owed to the people I was nominally supposed to be teaching. As the great Rabbi Chanina remarked two thousand years ago, "I have learned much from my teachers, more from my colleagues, and the most from my students." So, to my former students Drs. Evan Bierman, Nathan Miller, Stephanie Moyerman, Darcy Barron, Jonathan Kaufman, Frederick Matsuda, Dave Boettger, Chang Feng, Praween Siritanasak, and especially to my current students Logan Howe, Lindsay Lowry, Marty Navaroli, David Leon, and Tucker Elleflot: I owe you much. Thank you for enduring a sometimes absentee advisor while he was busy writing a book about being a good mentor.

This book owes its existence to my brother-from-another-mother, Stephon Alexander. Nothing has changed between us since those late-night intellectual jam sessions spent together as Nobel hopefuls at Brown University twenty-five years ago. Then, as now, Stephon's encouragement and wisdom kept me going through the darkness. Thank you for believing that mine was a story that needed to be told.

To my three best friends, my brothers, Kevin, Nick, and Shaya: we've been through everything together and our bonds will never break. You don't get to choose your brothers, but if I had been given the choice, I would surely choose you. I love you more than you will ever know.

Sean Carroll, Sara Lippincott, Max and John Brockman, and Bruce Lieberman helped me from the outset and continued to assist during the twisty journey from there to here. Thanks, too, to T. J. Kelleher for encouraging me to tell a different kind of scientific story, less victory lap than autobiographical account of how science actually gets done.

Liz Kruesi was an astronomical angel, finding holes and helping to patch them right back up. Thanks for your brilliant insights and words of encouragement when I needed them most. Thanks as well to Sarah Scoles for connecting me to Liz and for her own prose that continues to inspire me.

I owe much gratitude for sage advice, wisdom, and loving support, which came in ample supply, from my in-laws Robert and Allison Price, Jim and Rebecca Brewer, and David Price.

Dennis Prager, Kam Arnold, Eric Ertel, Sean Levi, Yoel Rephaeli, Todd Salovey, Robert DeLaurentis, Debra Kellner, Ido Ben Dayan, Lucian Iancovici, David Pinn, Greg Anderson, Andy Friedman, Raphael Flauger, and Rae Armantrout were so generous with their time and friendship that I am embarrassed by my inability to repay them. Naomi and Jay Pasachoff were the dynamic duo that set me off on a career as a thirteen-year-old and they continue to inspire and improve my scholarship. Thanks to Grant Teply, Andrew Jaffe, and Robert Caldwell for critical fact, logic, and grammatical oversight.

Lyman Page, David Spergel, Paul Steinhardt, Helge Kragh, Martin Rees, and Roger Penrose gave me vital insights and input as this book came to a conclusion. I thank you all for spending so much time with me and my ideas. David Brin did his best to transform me into a reader's writer, illuminating the opaque parts of my writing, which were numerous. I cannot presume that I achieved the lofty goals he set for me. But I tried.

Marilyn and Jim Simons have been there for me since before I was born. I owe you both so much. You are my role models, my connection to my roots, and I'm honored to have you as friends.

Just saying the name Allegra Huston brings me joy. She was my ace, coming in to rescue me long before it was too late. She helped me craft the book into something better than I ever dreamed possible. For your encouragement, insight, and acumen I am eternally grateful.

Some artists torture themselves. Instead it was I who tortured the brilliant artist named Shaffer Grubb, who tirelessly helped me bring complex visualizations to light.

My three academic advisors, Peter Timbie, Phil Lubin, and Alex Polnarev, taught me what it means to be a scientist and a mentor. I don't take for granted that I've achieved the lofty heights of their mentorship, but I hope I have shown myself worthy of the many thankless hours they spent with me over the decades.

My friends and colleagues from my Caltech days all played a role too fundamental to fully capture, so let me just say to Marc Kamionkowski, Ravinder Batia, Sunil Golwala, Yonah Solomon, and Kathy Deniston: I owe you so much. Thank you for being there in one of the most wonderful and interesting times in my life.

I never thought I'd have a spiritual advisor, but Rabbi Jeff Wohlgelernter probably never thought he'd acquire me as a student either. His wisdom, generosity, humanity, and brilliant insights into life make me want to be a better man.

Susan Sanfrey, Erin Lovett, Steve Colca, and Anna Oler are among the consummate professionals at Norton who made this book come to life.

Jed Peterson provided desperately needed feedback during a critical phase in the writing of the book. I hope he will find it worthy of checking out from his local library.

Jeff Shreve is a throwback to the golden age of literary editors: a tireless professional with wisdom far beyond his years. His guidance, support, and skillful editing were invaluable. Thank you for helping a neophyte navigate the shoals of publishing and for keeping me sane. You made this book what it is.

It's said that some women are born great, while some have greatness thrust upon them. Quynh Do represents more than a little of both, having been conscripted at the most fraught moment, when the prospect of completing this book appeared completely lost. She gave me reason for hope, and for that I am deeply grateful.

Peggy McCoy, Sheldon Brown, Hans Paar, Jonah Saidan, Erik Viirre, Patrick Coleman, Daniel Eden, Mark Thiemens, George Fuller, and Lynda Claasson at UC San Diego provided essential support, advice, and assistance exactly when I needed it. Thank you for making our university the remarkably vibrant place of learning it is.

Ray Keating deserves my gratitude for providing me my love of flight, and a home and the clothes on my back for more than a decade.

In an ever-decreasing federal funding world, agencies such as NASA, the Department of Energy, and the National Science Foundation are more heroic and vital than ever. Their work is crucial to the health of the nation and they are deeply appreciated. Thanks to the modern-day Medici, the Simons and Heising–Simons Foundations, for supporting the next chapter of cosmological research.

To all my BICEP2 teammates: Peter Ade, Randol Aikin, Denis Barkats, Steve Benton, Evan Bierman, Colin Bischoff, Jamie Bock, Justus Brevik, Immanuel Buder, Eric Bullock, Cynthia Chiang, Darren Dowell, Lionel Duband, Jeff Filippini, Stefan Fliescher, Sunil Golwala, Mark Halpern, Matt Hasselfield, Sergei Hildebrandt, Gene Hilton, Viktor Hristov, Kent Irwin, Kirit Karkare, Sarah Kernasovskiy, John Kovac, Chao-Lin Kuo, Erik Leitch, Martin Lueker, Pete Mason, Bart Netterfield, Hien Nguyen, Roger O'Brient, Walt Ogburn, Angiola Orlando, Clem Pryke, Carl Reintsema, Steffen Richter, Robert Schwarz, Chris Sheehy, Zak Staniszewski, Rashmi Sudiwala, Yuki Takahashi, Grant Teply, Jamie Tolan, Anthony Turner, Abby Vieregg, Chin Lin Wong, and Ki Won Yoon, all of you amazing scientists who froze, bled, sweated, and cried so that BICEP2 could come to life: thank you. You accomplished something spectacular and for that you should forever be proud. Of course, missing from this illustrious cast of brilliant scientists is our mentor, colleague, and friend, Andrew Lange, a man to whom much is owed.

My Simons Observatory collaborators have become a wonderful new tribe that inspire me. I only hope to match the passion, brilliance, and dedication you all show every day: Peter Ade, Zeeshan Ahmed, Simone Aiola, James Aguirre, Rupert Allison, David Alonso, Marcello Alvarez, Kam Arnold, Jason Austermann, Humna Awan, Carlo Baccigalupi, Taylor Baildon, Darcy Barron, Nick Battaglia, Richard Battye, Eric Baxter, James Beall, Shawn Beckman, Rachel Bean, Fedri Bianchini, Stephen Boada, Dick Bond, Julian Borril, Mike Brown, Sarah Marie Bruno, Sean Bryan, Erminia Calabrese, Paolo Calisse, Julien Carron, Anthony Challinor, Yuji Chinone, Hsiao-Mei Sherry Cho, Steve Choi, Devin Crichton, Joanne Cohn, Carlo Contaldi, Will Coulson, Nicholas Cothard, Kevin Crowley, Ari Cukierman, Mark Devlin, Simon Dicker, Joy Didier, Matt Dobbs, Bradley Dober, Scott Dodelson, Shannon Duff, Adri Duivenvoord, Jo Dunkley, John Dustako, Josquin Errard, Alex van Engelen, Giulio Fabbian, Raphael Flauger, Stephen Feeney, Simone Ferraro, Pedro Fluxa, Katie Freese, Josef Frisch, Andrei Frolov, George Fuller, Brittany Fuzia, Nicholas Galitzki, Patricio Gallardo, Jiansong Gao, Eric Gawiser, Martina Gerbino, Jose Tomás Galvez Gehrsi, Vera Gluscevic, Neil Goeckner-Wald, Joey Golec, Sam Gordon, Megan Gralla, Dan Green, Arpi Grigorian, John Groh, Chris Groppi, Jon Gudmundsson, Tijmen de Haan, Mark Halpern, Peter Hargrave, Masaya Hasegawa, Makoto Hattori, Hasashi Hazumi, Shawn Henderson, Carlos Hervias Caimapo, Charlie Hill, Colin Hill, Gene Hilton, Matt Hilton, Adam Hincks, Gary Hinshaw, Renee Hlozek, Shirley Ho, Shuay-Pwu Ho, Dongwon Hon, Logan Howe, Ziqi Huang, Johannes Hubmayr, Kevin Huffenberg, Jack Hughes, Anna Ijjas, Margaret Iakpe, Kent Irwin, Bhuvnesh Jain, Andrew Jaffe, Daisuke Kaneko, Ethan Karpel, Nobuhiko Katayama, Sarah Kernasovskiy, Reijo Keskitalo. Theodore Kisner, Jeff Klein, Kenda Knowles, Brian Koopman, Arthur Kosowsky,

Nicoletta Krachmalnicoff, Stephen Kuenstner, Chao-Lin Kuo, Akito Kusaka, Jeff Van Lanen, Adrian Lee, David Leon, Danielle Leonard, Jason Leung, Anthony Lewis, Yaqiong Li, Michele Limon, Eric Linder, Jia Liu, Carlos Lopez Caraballo, Thibaut Louis, Lindsay Lowry, Marius Lungu, Mat Madhavacheril, Daisy Mak, Hamdi Mani, Ben Mates, Fred Matsuda, Loïc Maurin, Phil Mauskopf, Andrew May, Chris McKenny, Jeff McMahon, Daniel Meerburg, Joel Meyers, Amber Miller, Mark Mirmelstein, Kavi Moodley, Sigurd Naess, Martin Navaroli, Laura Newburgh, Mike Niemack, Haruki Nishoni, Jack Orlowski, Lyman Page, Bruce Partridge, Enzo Pascale, Julien Peloton, Francesca Perrotta, Lucio Piccrillio, Rolando Planella, Davide Poletti, Heather Prince, Robert Puddu, Giuseppe Puglis, Aisha Quadir, Srinivasan Raghunathan, Chris Raum, Christian Reichardt, Yoel Rephaeli, Dominik Reichers, Felipe Rojas, Aditya Rotti, Sharon Sadeh, Emmanuel Schaan, Neelima Sehgal, Uroš Seljak, Blake Sherwin, Meir Shimon, Jon Sievers, Precious Sikhosana, Max Silva Feaver, Sara Simon, Adrian Sinclair, Praween Siritanasak, Steve Smith, David Spergel, Suzanne Staggs, George Stein, Jason Stevens, Cristobal Sifon, Radek Stompor, Rashmi Sudiwala, Aritoki Suzuki, Osamu Tajima, Satoru Takakura, Grant Teply, Dan Thomas, Bob Thornton, Hy Trac, Joel Ullom, Sunny Vagnozi, Eve Vavagiakis, Michael Vissers, Kasey Wagoner, John Ward, Ben Westbrook, Nathan Whitehorn, Dan van Winkle, Zhilei Xu, Fernando Zago, and Ningfeng Zhi.

There is no comparison between the effort this book required and the nearly infinite expenditure of kindness, love, and late-night proofreading provided by my darling wife, Sarah. Long after the last embers of the cosmos have vanished, my love for you will endure. This book couldn't have happened without you, Sarah. You have burdened me with blessings. You are my universe.

NOTES

Introduction
A NOBLE WILL

1 "Full text of Alfred Nobel's Will," https://www.nobelprize.org/alfred
_nobel/will/.

2 "John F. Kennedy Speeches: Remarks at the Convocation of the United
Negro College Fund, Indianapolis, Indiana, April 12, 1959," John F.
Kennedy Presidential Library and Museum, https://www.jfklibrary.org
/Research/Research-Aids/JFK-Speeches/Indianapolis-IN_19590412
.aspx.

3 The list of *nominators* for the prize in physiology or medicine is openly
published each year. See Magdalena Eriksson, "A Great Prize Ages with
Grace," *Science and Technology Perspective* 80, no. 4 (2002): 62–64.
Unlike the other Nobel Prizes, the Nobel Peace Prize makes public the
nominees under consideration.

4 Lise Meitner, along with Otto Hahn and Otto Frisch, discovered nuclear
fission. Yet Otto Hahn alone won the 1948 Nobel Prize in Chemistry for
this discovery.

5 Daniel Charles, *Between Genius and Genocide: The Tragedy of Fritz
Haber, Father of Chemical Warfare* (London: Pimlico, 2006).

6 Associated Press, "Desmond Tutu, 2 other Nobel Peace Prize laureates
contest 2012 winner choice of EU," *Fox News*, November 30, 2012, http://
www.foxnews.com/world/2012/11/30/desmond-tutu-2-other-nobel-
peace-prize-laureates-contest-2012-winner-choice-eu.html.

7 Valerie Richardson, "Critics mock Nobel committee for handing literature prize to Bob Dylan," *Washington Times*, October 13, 2016,

http://www.washingtontimes.com/news/2016/oct/13/bob-dylans-nobel-prize-sets-off-literature-debate/.

8 You can obtain your own at: https://spie.org/Documents/resources/Free%20Posters/copyright%20posters/Mayer-poster-delivery.pdf.

9 Some examples: Peter Doherty, *The Beginner's Guide to Winning the Nobel Prize: Advice for Young Scientists* (New York: Columbia University Press, 2006); David Carter, *How to Win the Nobel Prize in Literature* (London: Hesperus Press, 2012); Tony Goldsmith, *How to Win a Nobel Prize* (independently published, 2017).

10 Rebecca Hersher, "How Much Is a Nobel Prize Medal Worth?" *The Two-Way: Breaking News from NPR*, October 16, 2016, http://www.npr.org/sections/thetwo-way/2016/10/16/498146211/how-much-is-a-nobel-prize-medal-worth.

Chapter 2
LOSING MY RELIGIONS

1 Galileo Galilei, letter to Madame Christina of Lorraine, Grand Duchess of Tuscany, 1615, in *Discoveries and Opinions of Galileo*, translated and edited by Stillman Drake (New York: Doubleday Anchor, 1957).

2 J. Ax, "Group-Theoretic Treatment of the Axioms of Quantum Mechanics," *Foundations of Physics* 6, no. 4 (1976): 371–99; and J. Ax, "The Elementary Foundations of Spacetime," *Foundations of Physics* 8, no. 7–8 (1978): 507, https://doi.org/10.1007/BF00717578.

3 Andrew H. Jaffe, Mark Kamionkowski, and Limin Wang, "Polarization pursuers' guide," *Physical Review D* 61, 083501 (2000), https://journals.aps.org/prd/abstract/10.1103/PhysRevD.61.083501.

4 Brian Keating, "A search for the large angular scale polarization of the cosmic microwave background," PhD thesis, Brown University, 2000, http://adsabs.harvard.edu/abs/2000PhDT.......176K.

5 See Wikipedia, https://en.wikipedia.org/wiki/BOOMERanG_experiment. The name was also a play on words, for the balloon was to set out from and return to the same place on the coast of Antarctica during the two-week period when the South Polar Vortex sets up each year in the austral atmosphere.

6 "In Memoriam: Andrew Lange," California Institute of Technology website, http://calteches.library.caltech.edu/4340/1/Memoriam.pdf.

7 A. Miller et al., "A Measurement of the Angular Power Spectrum of the CMB from l = 100 to 400," *Astrophysical Journal* 524 (1999): L1–L4, https://doi.org/10.1086/312293.

Chapter 3
A BRIEF HISTORY OF TIME MACHINES

1 Mario Biagioli, *Galileo's Instruments of Credit: Telescopes, Images, Secrecy* (Chicago: University of Chicago Press, 2006).

2 Astonishingly, Simon Marius discovered (and documented) the moons of Jupiter *one* day after Galileo did. See J. M. Pasachoff, "Simon Marius's Mundus Iovialis: 400th Anniversary in Galileo's Shadow," *Journal for the History of Astronomy* 46, no. 2 (2015): 218–34, http://journals .sagepub.com/doi/full/10.1177/0021828615585493.

3 Galileo Galilei, *Sidereus Nuncius* (Venice: Thomas Baglioni, 1610). Astrology's outsized influence on astronomy is still felt. In the Romance languages, the days of the week are named after the Sun, the Moon, and the gods governing the peripatetic naked-eye planets.

4 Historian of astronomy Owen Gingerich has suggested that Copernicus was aware of the earlier heliocentric ideas of Aristarchus of Samos. See O. Gingerich, "Did Copernicus Owe a Debt to Aristarchus?" *Journal for the History of Astronomy* 16, no. 1 (1985): 37, http://adsabs.harvard.edu /full/1985JHA....16...37G.

5 In fact, Galileo would later propose the Jovian lunar system as history's first clock capable of accurately determining longitude on Earth.

6 Proof of the Sun's centrality would come with the discovery of stellar aberration by British astronomer James Bradley in 1728.

7 Even to the naked eye, let alone to a telescope, the Orion nebula is almost completely free of stars. Some scholars suggest that Galileo intentionally avoided discussing the Orion nebula, perhaps the most prominent nebula there is, because he "did not wish to vitiate his argument" that nebulae were made of stars alone. See Galileo Galilei, *Sidereus Nuncius, or The Sidereal Messenger,* translated by Albert Van Helden (Chicago: University of Chicago Press, 1989). Galileo deliberately excluded the nebula in the sword of Orion, a conspicuous absence since it is visible to the naked eye. See Thomas G. Harrison, "The Orion Nebula: Where in History Is It?" *Quarterly Journal of the Royal Astronomical Society* 25 (1984): 65–79, http://adsabs.harvard .edu/full/1984QJRAS..25...65H. See also Owen Gingerich, "The Mysterious Nebulae, 1610–1924," *Journal of the Royal Astronomical Society of Canada* 81, no. 4 (1987): 113–27, http://adsabs.harvard.edu /full/1987JRASC..81..113G.

8 National Science Foundation, *Science and Engineering Indicators 2014,* chapter 7, "Science and Technology: Public Attitudes and Understand-

ing," p. 7-23, https://www.nsf.gov/statistics/seind14/content/chapter-7
/chapter-7.pdf.

9   Claire Brock, *The Comet Sweeper: Caroline Herschel's Astronomical
Ambition* (London: Icon, 2007).

10 F. G. W. Struve, *Etudes d'Astronomie Stellaire* (Paris: l'Académie
Impériale des Sciences, 1847).

11 Henrietta Swan Leavitt and Edward C. Pickering, "Periods of 25 Variable Stars in the Small Magellanic Cloud," *Harvard College Observatory
Circular* 173 (1912): 1–3, http://adsabs.harvard.edu/abs/1912HarCi.173...
.1L.

12 Harlow Shapley and Heber D. Curtis, "The Scale of the Universe,"
Bulletin of the National Research Council 2, pt. 3, no. 11 (1921): 217,
http://adsabs.harvard.edu/abs/1921BuNRC...2..171S.

Chapter 4
THE BIGGER THE BANG, THE BIGGER THE PROBLEMS

1   V. M. Slipher, "Nebulae," *Proceedings of the American Philosophical Society* 56 (1917): 403–9, http://adsabs.harvard.edu/abs/1917
PAPhS..56..403S.

2   Barbara Wolff, "The Nobel Prize in Physics 1921: What Happened to
the Prize Money?" Albert Einstein in the World Wide Web, March 2016,
www.einstein-website.de/z_information/nobelprizemoney.html.

3   E. Hubble, "A Relation Between Distance and Radial Velocity
among Extra-Galactic Nebulae," *Proceedings of the National Academy of Sciences* 15, no. 3 (1929): 168–73, http://adsabs.harvard.edu
/abs/1929PNAS...15..168H.

4   Cormac O'Raifeartaigh et al., "Einstein's steady-state theory: an abandoned model of the cosmos," https://arxiv.org/pdf/1402.0132.pdf.

5   There were other factors at work too, such as intrinsic biases associated with the photographic plates Hubble used. See David N. Spergel, Michael Bolte, and Wendy Freedman, "The age of the universe,"
Proceedings of the National Academy of Sciences 94, no. 13 (1997):
6579–84, http://www.pnas.org/content/94/13/6579.full.

6   Dominique Lambert, *The Atom of the Universe: The Life and Work of
Georges Lemaître* (Kraków: Copernicus Center Press, 2015).

7   And, for you nuclear aficionados, a bit of beryllium and lithium.

8   In 1941 Andrew McKellar made a rough measurement of the temperature of cyanogen gas clouds in the Milky Way galaxy. McKellar's data
ruled out any background heat that was warmer than about 3 kelvin.
A difference of two kelvin (between Alpher and Herman's predicted

5 kelvin and McKellar's measured 3 kelvin) may not sound like much unless you know that the energy density of blackbody radiation—the kind Alpher and Herman were predicting—increases with the fourth power of the blackbody's temperature. Therefore, Alpher and Herman predicted over ten times more energy in their thermal background than that which was tolerable. Cosmically speaking, they were predicting a Miami summer during a blizzard in Anchorage.

9 Gamow's estimates varied erratically, too, though at least they went consistently downward, from 7 kelvin in 1953 to 6 kelvin in 1956.

10 Helge Kragh, *Cosmology and Controversy: The Historical Development of Two Theories of the Universe* (Princeton: Princeton University Press, 1999).

11 F. Hoyle, "On Nuclear Reactions Occurring in Very Hot STARS. I. the Synthesis of Elements from Carbon to Nickel," *Astrophysical Journal Supplement* 1 (1954): 121, http://adsabs.harvard.edu/abs/1954ApJS....1..121H.

12 C. W. Cook, W. A. Fowler, C. C. Lauritsen, and T. Lauritsen, "B^{12}, C^{12}, and the Red Giants," *Physical Review* 107, no. 2 (1957): 508.

13 Edwin E. Salpeter, "Fallacies in Astronomy and Medicine," *Reports on Progress in Physics* 68, no. 12 (2005): 2747, http://iopscience.iop.org/article/10.1088/0034-4885/68/12/R02/meta.

14 F. Hoyle and R. J. Tayler, "The Mystery of the Cosmic Helium Abundance," *Nature* 203 (1964): 1108–10, https://doi.org/10.1038/2031108a0.

15 Interestingly, Google (Alphabet) is currently designing balloons for low-cost, wide-area access to the modern Internet. See https://www.google.com/loon/.

16 John Oakes, "Interview with Arno Penzias and Robert Wilson," Evidence for Christianity website, May 5, 2005, evidenceforchristianity.org/interview-with-arno-penzias-and-robert-wilson.

17 E. A. Ohm, "Project Echo: Receiving System," *Bell System Technical Journal* 40, no. 4 (1961): 1065, https://archive.org/details/bstj40-4-1065.

18 You can listen to the noise converted to audible frequencies at: http://www.npr.org/templates/story/story.php?storyId=4655517.

19 You can explore Penzias and Wilson's pigeon trap at the Smithsonian Institution; see https://airandspace.si.edu/exhibitions/explore-the-universe/online/images/2001-5347.jpg.

20 "Arno Penzias and Robert Wilson: Bell Labs, Holmdel, NJ," Historic Sites, APS Physics website, American Physical Society, https://www.aps.org/programs/outreach/history/historicsites/penziaswilson.cfm.

21 R. H. Dicke, P. J. E. Peebles, P. G. Roll, and D. T. Wilkinson, "Cosmic Black-Body Radiation," *Astrophysical Journal* 142 (1965): 414–19, http://adsabs.harvard.edu/abs/1965ApJ...142..414D.

22 "Newly Discovered Radio Radiation May Provide a Clue to the Origin of the Universe," Bell Telephone Laboratories press release, May 23, 1965, https://media-bell-labs-com.s3.amazonaws.com/pages/20140518_1641/1965_BL_Press_Release_on_Radiation.pdf.

23 Dicke at al., "Cosmic Black-Body Radiation," http://adsabs.harvard.edu/abs/1965ApJ...142..414D.

24 Dicke et al. mention that, even if there was a singular origin (i.e., the Big Bang), it too "might have been extremely hot in the first stages."

25 See http://www.nobelprize.org/nobel_prizes/physics/laureates/1978/index.html.

Chapter 5
BROKEN LENS 1: THE NOBEL PRIZE'S CREDIT PROBLEM

1 See https://www.nobelprize.org/nobel_prizes/physics/laureates/2013/press.html.

2 G. S. Guralnik and C. R. Hagen, "Where Have All the Goldstone Bosons Gone?" https://arxiv.org/pdf/1401.6924.pdf.

3 "Gerald S. Guralnik, Chancellor's Professor of Physics," obituary, Brown University website, https://news.brown.edu/articles/2014/04/guralnik.

4 Margaret Burbidge, "Watcher of the Skies," *Annual Reviews of Astronomy and Astrophysics* 32 (1994): 1, http://adsbit.harvard.edu/cgi-bin/nph-iarticle_query?bibcode=1994ARA%26A..32....1B.

5 Sarah Scoles, "How Vera Rubin confirmed dark matter," *Astronomy*, October 4, 2016, http://www.astronomy.com/news/2016/10/vera-rubin.

6 Charles Seife, "Troubled by Glitches, Tevatron Scrambles to Retain Its Edge," *Science* 295, no. 5557 (2002): 36, http://science.sciencemag.org/content/295/5557/news-summaries.

7 Lisa Randall, "Why Vera Rubin deserved a Nobel," *New York Times*, January 4, 2017, https://www.nytimes.com/2017/01/04/opinion/why-vera-rubin-deserved-a-nobel.html?_r=0.

8 Dennis Overbye, "Vera Rubin, 88, Dies; Opened Doors in Astronomy, and for Women," *New York Times*, December 27, 2016, https://www.nytimes.com/2016/12/27/science/vera-rubin-astronomist-who-made-the-case-for-dark-matter-dies-at-88.html.

9 Adrian Cho, "Will Nobel Prize overlook master builder of gravitational wave detectors?" *Science*, September 27, 2016, http://www.sciencemag.org/news/2016/09/will-nobel-prize-overlook-master-builder-gravitational-wave-detectors.

10 "If they wait a year and give it to those three guys, at least I'll feel that

they thought about it. If they decide [to give it to them] this October [i.e., in 2016], I'll have more bad feelings because they won't have done their homework." Barry Barish, quoted ibid.

11 Immediately after the award was announced, the Nobel committee went out of its way to rationalize its decision not to award the 2017 prize to Drever posthumously. The description of the award for the public said that "Drever ultimately ended up outside the project's primary path." See "Cosmic Chirps," "The Nobel Prize in Physics 2017: Popular Science Background," Nobel Prize website, https://www.nobelprize.org/nobel_prizes/physics/laureates/2017/popular-physicsprize2017.pdf.

12 See http://www.nobelprize.org/nobel_organizations/nobelfoundation/statutes.html.

13 Zach Veilleux, "Nobel laureate Ralph Steinman dies at 68," Rockefeller University website, November 25, 2011, https://www.rockefeller.edu/news/1816-nobel-laureate-ralph-steinman-dies-at-68/.

14 "Ralph Steinman Remains Nobel Laureate," Nobel Foundation press release, October 3, 2011, https://www.nobelprize.org/nobel_organizations/nobelfoundation/press_releases_archive/2011/steinman.html.

15 Amy Davidson Sorkin, "Ralph Steinman: Death and the Nobel," *The New Yorker*, October 3, 2011, www.newyorker.com/news/amy-davidson/ralph-steinman-death-and-the-nobel.

16 See https://www.nobelprize.org/nomination/physics/.

17 See https://www.nobelprize.org/nobel_prizes/lists/laureates_ages/physics_ages.html.

Chapter 6
ASHES TO ASHES

1 Mike Wall, "Cosmic Inflation Theory Confirmed? Q&A with Robert Wilson, Co-Discoverer of Big Bang Echo," Space.com, March 17, 2014, www.space.com/25094-big-bang-inflation-cmb-wilson-interview.html.

2 Steven Weinberg, *Gravitation and Cosmology* (Hoboken, NJ: John Wiley & Sons, 1972).

3 Kim McDonald, "Renowned UC San Diego Astrophysicist and Astronomer Dies at 84," UC San Diego News Center, January 28, 2010, http://ucsdnews.ucsd.edu/archive/newsrel/science/01-10Burbidge.asp.

4 After the first stars, known as Population III, lived out their lives, they produced the raw materials needed to make successive generations of stars, known as Population II and Population I (the youngest category that comprises the bulk of stars in our galaxy).

5 G. Burbidge and F. Hoyle, "The Origin of Helium and the Other Light

Elements," *Astrophysical Journal* 509 (1998): L1–L3, http://iopscience
.iop.org/article/10.1086/311756/pdf.

6 Energy and temperature, though related, are not identical. A million-kilogram comet moving through space at ten kilometers per second has a lot of energy but could actually be only a few degrees above absolute zero.

7 Fred Hoyle, *Home Is Where the Wind Blows: Chapters from a Cosmologist's Life* (Oxford: Oxford University Press, 1994), 413.

8 Sunlight exhibits the same phenomenon. In the absence of significant scattering, e.g. at high noon, the Sun's light is greenish yellow, but at sunset it is reddish orange (diluted of its blue content) due to the dust in our atmosphere, which is thicker along the horizon.

9 R. J. Trumpler, "Absorption of Light in the Galactic System," *Publications of the Astronomical Society of the Pacific* 42, no. 248 (1930): 214, http://adsabs.harvard.edu/abs/1930PASP...42..214T; Jessie Rudnick, "On the Reddening in B-Type Stars," *Astrophysical Journal* 83 (1936): 394, http://adsabs.harvard.edu/cgi-bin/bib_query?1936ApJ....83..394R.

10 J. S. Hall, "Observations of the Polarized Light from Stars," *Science* 109, no. 2825 (1949): 166–67, http://science.sciencemag.org/content/109/2825/166.

11 Aigen Li, "Cosmic Needles versus Cosmic Microwave Background Radiation," *Astrophysical Journal* 584, no. 2 (2003): 593, http://iopscience
.iop.org/article/10.1086/345753/meta.

12 Leverett Davis, Jr., and Jesse L. Greenstein, "The Polarization of Starlight by Aligned Dust Grains," *Astrophysical Journal* 114 (1951): 206.

13 F. Hoyle and N. C. Wickramasinghe, "Dust in Supernova Explosions," *Nature* 226 (1970): 62–63, https://doi.org/10.1038/226062a0.

14 J. V. Narlikar et al., "Cosmic Iron Whiskers: Their Origin, Length Distribution and Astrophysical Consequences," *International Journal of Modern Physics D* 6 (1997): 125, http://www.worldscientific.com/doi/abs
/10.1142/S0218271897000108.

15 Jennifer Hackett, "How to Find Tiny Meteorites at Home," *Scientific American*, April 1, 2016, https://www.scientificamerican.com/article/how-to
-find-tiny-meteorites-at-home/.

16 Recently, the Cosmic Dust Analyzer on board the Cassini spacecraft revealed evidence for interstellar dust containing iron. See N. Altobelli et al., "Flux and Composition of Interstellar Dust at Saturn from Cassini's Cosmic Dust Analyzer," *Science* 352, no. 6283 (2016): 312–18, http://science.sciencemag.org/content/352/6283/312.

17 S. S. Brenner, "Metal Whiskers," *Scientific American* 203, no. 1 (1960): 64–72.

18 Anthony N. Aguirre, "Dust Versus Cosmic Acceleration," *Astrophysical Journal Letters* 512, no. 1 (1999), http://iopscience.iop.org/article
/10.1086/311862.

19 Glenn Roberts, Jr., "New Study Maps Space Dust in 3-D," Berkeley Lab website, March 22, 2017, http://newscenter.lbl.gov/2017/03/22/new-study-maps-space-dust-in-3-d/.

20 G. Burbidge, "Quasi-Steady State Cosmology," Proceedings of Frontiers of the Universe Conference, June 17–23, 2001, https://arxiv.org/abs/astro-ph/0108051.

21 J. C. Mather et al., "A preliminary measurement of the cosmic microwave background spectrum by the Cosmic Background Explorer (COBE) satellite," *Astrophysical Journal (Part 2 – Letters)* 354 (1990): L37–L40, http://adsabs.harvard.edu/abs/1990ApJ...354L..37M.

22 Ironically, the first astronomer to win a Nobel Prize for a theoretical discovery was Hans Bethe, the physicist whose name Gamow, a known punster, had added (over Alpher's objections) to his and Alpher's 1948 paper outlining the Big Bang theory. Gamow added Bethe not because he had contributed in any way but because it amused Gamow that the three names sounded like the first three letters of the Greek alphabet (alpha, beta, gamma)—appropriate, in Gamow's mind at least, for a theory about the beginning of the universe. Fortunately for Alpher's sanity, Bethe's Nobel Prize was awarded for his work on energy production in stars, not for the Big Bang theory.

23 "The Crafoord Prize 1997," Royal Swedish Academy of Sciences press release, http://www.crafoordprize.se/press/arkivpressreleases/thecrafoordprize1997.5.32d4db7210df50fec2d800018163.html.

24 Burbidge, "Quasi-Steady State Cosmology," https://arxiv.org/abs/astro-ph/0108051.

25 J. Kovac et al., "Detection of Polarization in the Cosmic Microwave Background using DASI," *Nature* 420 (2002): 772–87, https://arxiv.org/abs/astro-ph/0209478.

26 Anthony N. Aguirre, "The Cosmic Background Radiation in a Cold Big Bang," *Astrophysical Journal* 533, no. 1 (2000): È18, http://iopscience.iop.org/article/10.1086/308660.

27 Jayant V. Narlikar and Geoffrey Burbidge, *Facts and Speculations in Cosmology* (Cambridge, UK: Cambridge University Press, 2009).

Chapter 7
THE SPARK THAT IGNITED THE BIG BANG

1 A. A. Penzias and R. W. Wilson, "A Measurement of Excess Antenna Temperature at 4080 Mc/s," *Astrophysical Journal* 142 (1965): 419–21, http://adsabs.harvard.edu/abs/1965ApJ...142..419P.

2 Quoted in Jane Gregory and Steven Miller, *Science In Public: Communication, Culture, and Credibility* (New York: Basic Books, 2000).

3 Stephon Alexander, *The Jazz of Physics: The Secret Link Between Music and the Structure of the Universe* (New York: Basic Books, 2016).

4 "Ned Wright's Cosmology Tutorial," UCLA website, http://www.astro.ucla.edu/~wright/cosmoall.htm.

5 Megan Vandre, "Before the Big Bang," *MIT Technology Review*, February 1, 2003, https://www.technologyreview.com/s/401796/before-the-big-bang/.

6 R. H. Dicke and P. J. E. Peebles, "The Big Bang Cosmology: Enigmas and Nostrums," in S. W. Hawking and W. Israel, eds., *General Relativity: An Einstein Centenary Survey* (Cambridge, UK: Cambridge University Press, 1979).

7 Vandre, "Before the Big Bang," https://www.technologyreview.com/s/401796/before-the-big-bang/.

8 Alan H. Guth, "Inflationary Universe: A possible solution to the horizon and flatness problems," *Physical Review D* 23, no. 2 (1981): 347, https://link.aps.org/doi/10.1103/PhysRevD.23.347.

9 Alan H. Guth, "Inflation," *Proceedings of the National Academy of Sciences* 90, no. 11 (1993): 4871–77, http://www.pnas.org/content/90/11/4871.short.

10 Arthur C. Clarke Center for Human Imagination, UC San Diego, *Endless Universe* podcast, episode 5: "Limits of Understanding," http://imagination.ucsd.edu/_wp/podcast/episode-5-limits-of-understanding/.

11 Steinhardt and Turner, along with physicist James Bardeen, came to a conclusion that was discovered independently by three separate groups: Stephen Hawking, Alexei Starobinsky, and Guth and his colleague So-Young Pi.

12 L. F. Abbott and M. B. Wise, "Constraints on Generalized Inflation Cosmologies," *Nuclear Physics B* 244, no. 2 (1984): 541, http://adsabs.harvard.edu/abs/1984NuPhB.244..541A, built upon the earlier pioneering work of Alexei Starobinsky, who not only came up with a model of inflation at the same time as Guth but also delineated the behavior of gravitational waves within the early universe. See A. Starobinsky, "Spectrum of Relict Gravitational Radiation and the Early State of the Universe," *Journal of Experimental and Theoretical Physics Letters* 30 (1979): 682, http://adsabs.harvard.edu/abs/1979ZhPmR..30..719S.

13 Martin Rees, "Polarization and Spectrum of the Primeval Radiation in an Anisotropic Universe," *Astrophysical Journal* 153 (1968): L1, http://adsabs.harvard.edu/abs/1968ApJ...153L...1R.

14 In the mid-1990s two groups of cosmologists, one composed of Marc Kamionkowski, Arthur Kosowsky, and Albert Stebbins, and the other

composed of Uroš Seljak and Matias Zaldarriaga, updated Polnarev's calculations and made them more "user friendly," at least for cosmological consumers like me.

Chapter 8
BICEP: THE ULTIMATE TIME MACHINE

1 Lawrence M. Krauss and Frank Wilczek, "From B Modes to Quantum Gravity and Unification of Forces," https://arxiv.org/abs/1404.0634.
2 B. Keating et al., "BICEP: a large angular scale CMB polarimeter," *Polarimetry in Astronomy: Proceedings of the SPIE* 4843 (2003): 284–95, http://adsabs.harvard.edu/abs/2003SPIE.4843..284K.
3 J. N. Tinsley et al., "Direct Detection of a Single Photon by Humans," *Nature Communications* 7 (2016): 12172, https://www.nature.com/articles/ncomms12172.
4 The noise in the denominator is sometimes referred to as "sigma," the universal symbol for uncertainty; thus, a detection with an SNR of three is "three sigma."
5 "In Memoriam: Andrew Lange."
6 "Wide bandwidth polarization modulator for microwave and mm-wavelengths," US patent 7501909 B2, https://www.google.com/patents/US7501909.

Chapter 9
HEROES OF FIRE, HEROES OF ICE

1 Both Curies suffered accidental and intentionally inflicted radiation burns; see Richard Mould, "Pierre Curie, 1859–1906," *Current Oncology* 14, no. 2 (2007): 74–82, http://www.current-oncology.com/index.php/oncology/article/view/126. Their laboratory notebooks are still radioactive; to study them you must sign a liability release and wear protective gear. See Barbara Tasch, "Personal effects of 'the mother of modern physics' will be radioactive for another 1500 years," *Business Insider*, August 24, 2015, http://www.businessinsider.com/marie-curie-radioactive-papers-2015-8.
2 "Doomed Expedition to the South Pole, 1912," EyeWitness to History.com, www.eyewitnesstohistory.com/scott.htm.
3 Susan Solomon, *The Coldest March* (New Haven: Yale University Press, 2002).
4 The 16PF and the MMPI-2 exams, used to determine psychological fitness, are available online.

5 A. Lazarian, "Grain Alignment and CMB Polarization Studies," 2008, https://arxiv.org/abs/0811.1020.

6 E. M. Bierman et al., "A Millimeter-Wave Galactic Plane Survey with the BICEP Polarimeter," *Astrophysical Journal* 741, no. 2 (2011): 81, http://iopscience.iop.org/article/10.1088/0004-637X/741/2/81/meta.

Chapter 10
BROKEN LENS 2: THE NOBEL PRIZE'S CASH PROBLEM

1 Ellie Bothwell, "Nobel laureate says scientific breakthrough 'would not be possible today,'" World University Rankings website of The Times Higher Education, September 29, 2016, https://www.times highereducation.com/news/nobel-laureate-says-scientific-breakthrough -would-not-be-possible-today.

2 Ibid.

3 Ibid.

4 "The Presidential Early Career Award for Scientists and Engineers: Recipient Details—Brian Keating," National Science Foundation web- site, https://www.nsf.gov/awards/PECASE/recip_details.jsp;jsessionid= 83E30069675B24B23A2117EB9E241D39?pecase_id=208.

5 See "Report to the National Science Board on the National Science Foundation's Merit Review Process, Fiscal Year 2013," National Science Foundation, 2014, https://www.nsf.gov/nsb/publications/2014/nsb1432 .pdf. In astronomy the odds are even lower; see "Mathematical and Physical Sciences (MPS) Funding Rates," National Science Foundation website, https://www.nsf.gov/funding/funding-rates.jsp?org=MPS.

6 Harriet Zuckerman, "Stratification in American Science," *Sociologi- cal Inquiry* 40, no. 2 (1970): 235–57, http://onlinelibrary.wiley.com/doi /10.1111/j.1475-682X.1970.tb01010.x/abstract.

7 Derek J. de Solla Price, *Little Science, Big Science* (New York: Colum- bia University Press, 1963).

8 Paul A. Samuelson, "Paul A. Samuelson," in W. Breit and B. T. Hirsch, eds., *Lives of the Laureates: Eighteen Nobel Economists* (Cambridge, MA: MIT Press, 2004), 49–64.

9 R. K. Merton, "The Matthew Effect in Science," *Science* 159, no. 3810 (1968): 56–63, http://science.sciencemag.org/content/159/3810/56.

10 Harriet Zuckerman, "Nobel Laureates in Science: Patterns of Produc- tivity, Collaboration, and Authorship," *American Sociological Review* 32, no. 3 (1967): 391–403, https://doi.org/10.2307/2091086.

11 Rita Devlin Marier, "How much is a Nobel worth? A lot more than the

prize money," Phys.org, October 2, 2011, http://phys.org/news/2011-10-nobel-worth-lot-prize-money.html.

12 See, e.g., "The Nobel Prize and RAND," RAND Corporation, http://www.rand.org/about/history/nobel.html; "Global recognition for groundbreaking discovery," Nokia Bell Labs, https://www.bell-labs.com/our-people/recognition/; and Jeremy Alder, "50 Universities with the Most Nobel Prize Winners," Best Masters Programs website, http://www.bestmastersprograms.org/50-universities-with-the-most-nobel-prize-winners/.

13 U.S. Department of Energy, Office of Science, "Honors & Awards: DOE Nobel Laureates," https://science.energy.gov/about/honors-and-awards/doe-nobel-laureates/; Saul Perlmutter, "Supernovae, Dark Energy and the Accelerating Universe: How DOE Helped to Win (yet another) Nobel Prize," lecture at Lawrence Berkeley National Laboratory, January 13, 2012, https://www.osti.gov/scitech/biblio/1032838.

14 See https://www.nobelprize.org/nobel_organizations/nobel-center/.

15 See https://www.olympic.org/olympic-truce.

16 Chip Le Grand, "Rio 2016: price of success could be $9.2m per medal," *Australian*, August 2, 2016, http://www.theaustralian.com.au/news/nation/rio-2016-price-of-success-could-be-92m-per-medal/news-story/d0aa090e2413de0229963de361e902c9.

17 Annie Duke, "Would You Risk Death for an Olympic Medal? Temporal Discounting and the Zika Virus," *Huffpost* blog, August 5, 2017, www.huffingtonpost.com/annie-duke/would-you-risk-death-for_b_11331742.html. However, a more recent study comes to a different conclusion: see James Connor, Jules Woolf, and Jason Mazanov, "Would they dope? Revisiting the Goldman dilemma," *British Journal of Sports Medicine* 47, no. 11 (2013), http://bjsm.bmj.com/content/47/11/697.

18 Michael Linden, "Budget Cuts Set Funding Path to Historic Lows," Center for American Progress, January 29, 2013, https://www.americanprogress.org/issues/economy/reports/2013/01/29/50945/budget-cuts-set-funding-path-to-historic-lows/.

19 "NSF-funded LIGO pioneers named 2017 Nobel Prize in Physics laureates," National Science Foundation press release, October 3, 2017, https://www.nsf.gov/news/news_summ.jsp?cntn_id=243280; U.S. Department of Energy, Office of Science, "Honors and Awards: DOE Nobel Laureates"; "NIST and the Nobel: NIST Nobel Winners," National Institute of Standards and Technology, https://www.nist.gov/nist-and-nobel; and "Nobel Laureates," *The NIH Almanac*, National Institutes of Health, https://www.nih.gov/about-nih/what-we-do/nih-almanac/nobel-laureates.

20 Ronald J. Daniels, "A generation at risk: Young investigators and the

future of the biomedical workforce," *Proceedings of the National Academy of Sciences* 112, no. 2 (2014), www.pnas.org/content/112/2/313 .full.

21 Kevin Boudreau, Eva Catharina Guinan, Karim R. Lakhani, and Christoph Riedl, "The Novelty Paradox & Bias for Normal Science: Evidence from Randomized Medical Grant Proposal Evaluations," Harvard Business School Working Paper, No. 13–053, December 2012, https://dash .harvard.edu/handle/1/10001229.

22 David Callahan, "Inside the Simons Foundation: Big Philanthropy on the Frontiers of Science," Inside Philanthropy, May 17, 2016, http://www .insidephilanthropy.com/home/2016/5/17/inside-the-simons-foundation -big-philanthropy-on-the-frontie.html.

23 "Research Institutions Received Over $2.3 Billion in Private Funding for Basic Science in 2016," Science Philanthropy Alliance press release, February 13, 2017, http://www.sciencephilanthropyalliance.org/what- we-do/news/2-3-billion-in-private-funding-for-basic-science-in-2016/.

24 William J. Broad, "Billionaires with Big Ideas Are Privatizing American Science," *New York Times*, March 15, 2014, https://www.nytimes .com/2014/03/16/science/billionaires-with-big-ideas-are-privatizing -american-science.html?_r=0.

25 Ibid.

26 Lawrence M. Krauss, "Do the New, Big-Money Science Prizes Work?" *The New Yorker*, February 3, 2016, http://www.newyorker.com/tech/ elements/do-the-new-big-money-science-prizes-work.

27 "The birth of the web," CERN, https://home.cern/topics/birth-web.

28 Jonah Kanner and Alan Weinstein, "The Astrophysicists Who Faked It," *Nautilus*, November 3, 2016, http://nautil.us/issue/42/fakes/the -cosmologists-who-faked-it.

29 Shahram Heshmat, "What Is Confirmation Bias?" *Psychology Today*, April 23, 2015, https://www.psychologytoday.com/blog/science-choice /201504/what-is-confirmation-bias.

Chapter 11
ELATION!

1 G. Efstathiou and S. Gratton, "B-mode detection with an extended planck mission," *Journal of Cosmology and Astroparticle Physics* 2009 (June 2009), http://iopscience.iop.org/article/10.1088/1475-7516/ 2009/06/011.

2 J. Weber, "Evidence for Discovery of Gravitational Radiation," *Physi-*

cal Review Letters 22 (1969): 1320, https://journals.aps.org/prl/abstract /10.1103/PhysRevLett.22.1320.

3 Janna Levin, *Black Hole Blues and Other Songs from Outer Space* (New York: Knopf, 2016).

4 "OPERA experiment reports anomaly in flight time of neutrinos from CERN to Gran Sasso," CERN press release, September 23, 2011, updated June 8, 2012, https://press.cern/press-releases/2011/09/opera-experiment-reports-anomaly-flight-time-neutrinos-cern-gran-sasso. See also OPERA collaboration, "Measurement of the neutrino velocity with the OPERA detector in the CNGS beam," *Journal of High Energy Physics* 2012:93, http://link.springer.com/article/10.1007%2FJ HEP10%282012%29093.

5 Eugenie Samuel Reich, "Embattled neutrino project leaders step down," *Nature* News, April 2, 2012, https://www.nature.com/news/embattled -neutrino-project-leaders-step-down-1.10371.

6 Govert Schilling, *Ripples in Spacetime: Einstein, Gravitational Waves, and the Future of Astronomy* (Cambridge, MA: Belknap Press, 2017).

7 The original BICEP2 paper (P. A. R. Ade et al., "BICEP2 I: Detection of *B*-mode Polarization at Degree Angular Scales," March 17, 2014, https:// arxiv.org/pdf/1403.3985v1.pdf) cited the PowerPoint slides used by J.-Ph. Bernard (https://www.cosmos.esa.int:documents:387566:428323:4 7ESLAB_April_04_11_25_Bernard.pdf/) at the 2013 conference "47th ESLAB: The universe as seen by Planck" (European Space Agency website, https://www.cosmos.esa.int/web/planck/47-eslab).

8 Schilling, *Ripples in Spacetime.*

9 Ron Cowen, "How astronomers saw gravitational waves from the Big Bang," *Nature* News, March 17, 2014, http://www.nature.com/news/how -astronomers-saw-gravitational-waves-from-the-big-bang-1.14885.

10 Penzias and Wilson, "A Measurement of Excess Antenna Temperature at 4080 Mc/s," http://adsabs.harvard.edu/abs/1965ApJ...142..419P.

11 "FAQs," BICEP and Keck Array CMB Experiments official website, http://bicepkeck.org.

12 Keating et al., "BICEP," http://adsabs.harvard.edu/abs/2003SPIE .4843..284K.

13 Diana Steele, "On the Coverage of BICEP2," *ScienceWriters*, July 23, 2015, https://www.nasw.org/article/sciencewriters-coverage-bicep2.

14 Lisa Grossman, "First glimpse of big bang ripples from the universe's birth," *New Scientist*, March 17, 2014, https://www.newscientist.com /article/dn25235-first-glimpse-of-big-bang-ripples-from-universes-birth/.

15 Max Tegmark, "Good Morning, Inflation! Hello, Multiverse!"

Huffpost blog, March 17, 2014, http://www.huffingtonpost.com/max-tegmark/good-morning-inflation-he_b_4976707.html.

16 "NSF-funded researchers say Antarctic telescope may have provided the first direct evidence of cosmic inflation and the origins of the universe," National Science Foundation press release, March 17, 2014, https://www.nsf.gov/news/news_summ.jsp?cntn_id=130760.

17 "Stanford Professor Andrei Linde celebrates physics breakthrough," video, March 17, 2014, https://www.youtube.com/watch?v=ZlfIVEy_YOA.

18 Dennis Overbye, "Space Ripples Reveal Big Bang's Smoking Gun," *New York Times*, March 17, 2014, https://www.nytimes.com/2014/03/18/science/space/detection-of-waves-in-space-buttresses-landmark-theory-of-big-bang.html?hp&_r=0.

19 Lawrence M. Krauss, "A Scientific Breakthrough Lets Us See to the Beginning of Time," *The New Yorker*, March 17, 2014, http://www.newyorker.com/tech/elements/a-scientific-breakthrough-lets-us-see-to-the-beginning-of-time.

Chapter 12
INFLATION AND ITS DISCONTENTS

1 Milton Friedman, "Inflation and Unemployment," Nobel Memorial Lecture, December 13, 1976, http://www.nobelprize.org/nobel_prizes/economic-sciences/laureates/1976/friedman-lecture.pdf.

2 Allan Adams, "The discovery that could rewrite physics," TED talk, March 2014, https://www.ted.com/talks/allan_adams_the_discovery_that_could_rewrite_physics.

3 Brian Greene, "John Kovac," in "The 100 Most Influential People," *Time*, April 23, 2014, http://time.com/collection-post/70868/john-kovac-2014-time-100/.

4 Paul Steinhardt, "Natural Inflation," in G. W. Gibbons, S. W. Hawking, and S. T. C. Siklos, eds., *The Very Early Universe: Proceedings of the Nuffield Workshop, held at Cambridge, England, 21 June–9 July, 1982* (Cambridge, UK: Cambridge University Press, 1982); Alexander Vilenkin, "Birth of Inflationary Universes," *Physical Review D* 27, no. 12 (1983): 2848–55, https://doi.org/10.1103/PhysRevD.27.2848; and A. D. Linde, "Eternally Existing Self-Reproducing Chaotic Inflationary Universe," *Physics Letters B* 175, no. 4 (1986): 395–400, https://web.stanford.edu/~alinde/Eternal86.pdf.

5 Gabriel Popkin, "Swirling Bacteria Linked to the Physics of Phase Transi-

tions," *Quanta*, May 4, 2017, https://www.quantamagazine.org/swirling-bacteria-linked-to-the-physics-of-phase-transitions-20170504.

6 Susan Brown, "Colonies of Bacteria Fight for Resources with Lethal Protein," UC San Diego News Center, March 23, 2010, http://ucsdnews.ucsd.edu/archive/newsrel/science/03-23LethalProtein.asp.

7 Guth, "Inflation," http://www.pnas.org/content/90/11/4871.short.

8 Miriam Kramer, "Our Universe May Exist in a Multiverse, Cosmic Inflation Discovery Suggests," Space.com, March 18, 2014, https://www.space.com/25100-multiverse-cosmic-inflation-gravitational-waves.html.

9 For example, according to Stephen M. Feeney, Matthew C. Johnson, Daniel J. Mortlock, and Hiranya V. Peiris ("First Observational Tests of Eternal Inflation," *Physical Review Letters* 107, 071301 [2011], https://arxiv.org/abs/1012.1995), if a neighboring bubble universe bumped into our own, it would leave a distinct imprint on the cosmic microwave background. So far, there's been no evidence of such a "bruise" on the CMB. Another possible way to "see" the multiverse could come courtesy of black holes. In 2015, Alexander Vilenkin and his colleagues suggested that inflation would produce a unique distribution of black holes with various masses and sizes. By carefully measuring the distribution of black holes, using gravitational-wave detectors such as LIGO, they could provide evidence for inflation, and therefore the multiverse. See Marcus Woo, "Why the Multiverse Isn't Just Madness," *Science Friday*, January 26, 2017, https://www.sciencefriday.com/articles/why-the-multiverse-isnt-just-madness/.

10 Brandon Carter, "Large Number Coincidences and the Anthropic Principle in Cosmology," in *Confrontation of Cosmological Theories with Observational Data: Proceedings of the Symposium, Krakow, Poland, September 10–12, 1973* (Dordrecht: D. Reidel, 1974), 291–98.

11 He called it "the world ensemble philosophy."

12 Steven Weinberg, "Physics: What We Do and Don't Know," *New York Review of Books*, November 7, 2013, http://www.nybooks.com/articles/2013/11/07/physics-what-we-do-and-dont-know/.

13 Roger Penrose, "Faith, Fashion and Fantasy in the New Physics of the Universe," Louis Clark Vanuxem Lectures, Princeton University, October 17, 20, and 22, 2003, http://lectures.princeton.edu/2006/roger-penrose/.

14 Robert Crittenden et al., "Imprint of Gravitational Waves on the Cosmic Microwave Background," *Physical Review Letters* 71, no. 3 (1993): 324, https://doi.org/10.1103/PhysRevLett.71.324.

15 "Dirac Medallists 2002," International Centre for Theoretical Physics,

https://www.ictp.it/about-ictp/prizes-awards/the-dirac-medal/the-medal lists/dirac-medallists-2002.aspx.

16 Maggie McKee, "Ingenious: Paul J. Steinhardt," *Nautilus*, September 25, 2014, nautil.us/issue/17/big-bangs/ingenious-paul-j-steinhardt.

17 Paul J. Steinhardt, "2014: What Scientific Idea Is Ready for Retirement?" *Edge*, https://www.edge.org/response-detail/25405.

18 Karl. R. Popper, *Conjectures and Refutations: The Growth of Scientific Knowledge* (New York and London: Basic Books, 1962).

19 "This does not mean that Freud and Adler were not seeing certain things correctly; I personally do not doubt that much of what they say is of considerable importance, and may well play its part one day in a psychological science which is testable. But it does mean that those 'clinical observations' which analysts naïvely believe confirm their theory cannot do this any more than the daily confirmations which astrologers find in their practice." Popper, *Conjectures and Refutations*.

20 Helge Kragh, "The Criteria of Science, Cosmology and the Lessons of History," in Michael Heller, Bartosz Brozek, and Łukasz Kurek, eds., *Between Philosophy and Science* (Kraków: Copernicus Center Press, 2013). Kragh points out that even as late as 1982 Popper extolled the Steady State theory, calling it "a very fine and promising theory" and saying, "What was a metaphysical idea yesterday can become a testable theory tomorrow."

21 Natalie Wolchover and Peter Byrne, "In a Multiverse, What are the Odds?" *Quanta*, November 3, 2014, https://www.quantamagazine.org /20141103-in-a-multiverse-what-are-the-odds/.

22 Anna Ijjas and Paul J. Steinhardt, "Classically Stable Nonsingular Cosmological Bounces," *Physical Review Letters* 117, 121304 (2016), https://doi.org/10.1103/PhysRevLett.117.121304.

23 Max Tegmark, *Our Mathematical Universe* (New York: Knopf, 2014).

24 Tegmark, "Good Morning, Inflation!" http://www.huffingtonpost.com /max-tegmark/good-morning-inflation-he_b_4976707.html.

25. Amanda Gefter, "From Discovery to Dust," *Nova Next*, October 29, 2014, http://www.pbs.org/wgbh/nova/next/physics/bicep2/.

26 Dan Vergano, "Alan Guth: Waiting for the Big Bang," *National Geographic*, June 30, 2014, news.nationalgeographic.com/news/innovators/ 2014/06/140630-alan-guth-profile-inflation-cosmology-science/.

27 Ron Cowen, "Cosmology: Polar star," *Nature* News, March 31, 2014, http://www.nature.com/news/cosmology-polar-star-1.14954.

28 "Focus: Physicists Weigh in on BICEP2," *Physics* 7 (June 19, 2014): 65, https://physics.aps.org/articles/v7/65.

29 Hoyle, *Home Is Where the Wind Blows*, 413.

30 Maggie McKee, "Ingenious: Paul J. Steinhardt," nautil.us/issue/17/ big-bangs/ingenious-paul-j-steinhardt.

31 "Focus: Physicists Weigh in on BICEP2," https://physics.aps.org/articles /v7/65.

Chapter 13
BROKEN LENS 3: THE NOBEL PRIZE'S
COLLABORATION PROBLEM

1 Frank Close, *The Infinity Puzzle: Quantum Field Theory and the Hunt for an Orderly Universe* (Oxford: Oxford University Press, 2011).

2 A. Douglas Stone, "Fantasy Physics: Should Einstein have Won Seven Nobel Prizes," Princeton University Press blog, guest post, March 12, 2014, http://blog.press.princeton.edu/2014/03/12/fantasy-physics-should-einstein-have-won-seven-nobel-prizes/.

3 Philip Ball, "How 2 Pro-Nazi Nobelists Attacked Einstein's "Jewish Science," *Scientific American*, February 13, 2015, https://www.scientific american.com/article/how-2-pro-nazi-nobelists-attacked-einstein-s-jewish-science-excerpt1/.

4 de Solla Price, *Little Science, Big Science.*

5 Hub Zwart, "The Nobel Prize as a Reward Mechanism in the Genomics Era: Anonymous Researchers, Visible Managers and the Ethics of Excellence," *Journal of Bioethical Inquiry* 7, no. 3 (2010): 299–312, https://link.springer.com/article/10.1007/s11673-010-9248-0.

6 Bothwell, "Nobel laureate says scientific breakthrough 'would not be possible today,'" https://www.timeshighereducation.com/news/nobel-laureate-says-scientific-breakthrough-would-not-be-possible-today.

7 Ibid.

8 Zuckerman, "Nobel Laureates in Science: Patterns of Productivity, Collaboration, and Authorship," https://doi.org/10.2307/2091086.

9 "Vote for inflation Nobel Prize," *Vixra* (blog), https://vixra.wordpress .com/2014/03/20/vote-for-inflation-nobel-prize/.

10 P. E. Gibbs, "Who Might Get the Nobel Prize for Cosmic Inflation?" *Prespacetime Journal* 5, no. 3 (2014): 230–33, https://prespacetime.com/ index.php/pst/article/download/614/612.

11 Ibid.

12 https://www.nobelprize.org/podcast/.

13 Stuart Clark, "Gravitational waves gives Nobel prize committee another headache," *Guardian*, March 21, 2014, https://www.theguardian.com /science/2014/mar/21/gravitational-waves-nobel-prize-inflation.

14 Jim Al-Khalili, "Why the Nobel prizes need a shakeup," *Guardian*, October 8, 2012, https://www.theguardian.com/commentisfree/2012/oct/08/nobel-prizes-need-shakeup.

15 Harriet Zuckerman, *Scientific Elites: Nobel Laureates in the United States* (New Brunswick, NJ: Transaction, 1996).

16 Randy Kennedy, "Who Was That Food Stylist? Film Credits Roll On," *New York Times*, January 11, 2004, http://www.nytimes.com/2004/01/11/us/who-was-that-food-stylist-film-credits-roll-on.html?_r=0.

17 Elisabeth T. Crawford, *The Beginnings of the Nobel Institution: The Science Prizes, 1901–1915* (Cambridge, UK: Cambridge University Press, 1987).

Chapter 14
DEFLATION

1 Galileo Galilei, *Le operazioni del compasso geometrico e militare* (Venice: 1606).

2 Biagioli, *Galileo's Instruments of Credit*.

3 Hao Liu, Philipp Mertsch, and Subir Sarkar, "Fingerprints of Galactic Loop I on the Cosmic Microwave Background," *Astrophysics Journal* 789 (2014): L29, https://doi.org/ 10.1088/2041-8205/789/2/L29.

4 "WMAP, Planck, BICEP, and Inflation," lecture by David Spergel at "Cosmology after Planck" Colloquium, March 27, 2014, https://www.youtube.com/watch?v=j3fHkQa6818.

5 Raphael Flauger, J. Colin Hill, and David N. Spergel, "Toward an understanding of foreground emission in the BICEP2 region," *Journal of Cosmology and Astroparticle Physics* 2014:1405.7351, https://arxiv.org/abs/1405.7351.

6 Ron Cowen, "Gravitational wave discovery faces scrutiny," *Nature News*, May 16, 2014, http://www.nature.com/news/gravitational-wave-discovery-faces-scrutiny-1.15248; Dennis Overbye, "Astronomers Hedge on Big Bang Detection Claim," *New York Times*, June 19, 2014, https://www.nytimes.com/2014/06/20/science/space/scientists-debate-gravity-wave-detection-claim.html?_r=0.

7 Michael J. Mortonson and Uroš Seljak, "A joint analysis of Planck and BICEP2 B modes including dust polarization uncertainty," *Journal of Cosmology and Astroparticle Physics* 2014 (October 2014), http://iopscience.iop.org/article/10.1088/1475-7516/2014/10/035/meta;jsessionid=8590E71CAB59298BA35CDF6D3A5A6DC9.c1.iopscience.cld.iop.or.

8 Overbye, "Astronomers Hedge on Big Bang Detection Claim," https://

www.nytimes.com/2014/06/20/science/space/scientists-debate-gravity-wave-detection-claim.html?_r=0.

9 "Newly Discovered Radio Radiation May Provide a Clue to the Origin of the Universe," Bell Telephone Laboratories press release, May 23, 1965, https://media-bell-labs-com.s3.amazonaws.com/pages/20140518_1641/1965_BL_Press_Release_on_Radiation.pdf.

10 P. James E. Peebles, Lyman A. Page, Jr., and R. Bruce Partridge, eds., *Finding the Big Bang* (Cambridge, UK: Cambridge University Press, 2009).

11 Joel Achenbach, "Cosmic smash-up: BICEP2's big bang discovery getting dusted by new satellite data," *Achenblog* (blog), *Washington Post*, September 22, 2014, https://www.washingtonpost.com/news/achenblog/wp/2014/09/22/planck-satellite-shows-bicep2-telescope-make-have-seen-dust-not-the-big-bang/?utm_term=.a2caa5fe5d75.

12 Shannon Hall, "BICEP2 Was Wrong, But Publicly Sharing the Results Was Right," *The Crux* (blog), *Discover* website, January 30, 2015, blogs.discovermagazine.com/crux/2015/01/30/bicep2-wrong-sharing-results/#.WNMILhiZMmo.

13 Council for the Advancement of Science Writing, New Horizons in Science 2014 conference, Columbus, OH, October 19–20, 2014; see http://casw.org/conference-new-horizons/conference/new-horizons-science-2014.

14 Steele, "On the Coverage of BICEP2," https://www.nasw.org/article/sciencewriters-coverage-bicep2.

15 "Dust to Dust," editorial, *Nature* 514, no. 7522 (2014), http://www.nature.com/news/dust-to-dust-1.16137.

16 P. A. R. Ade et al., "Joint Analysis of BICEP2/*Keck Array* and *Planck* Data," *Physical Review Letters* 114, 101301 (2015), https://physics.aps.org/featured-article-pdf/10.1103/PhysRevLett.114.101301.

17 "Andrew Lange: USA: 2006 Balzan Prize for Observational Astronomy and Astrophysics," International Balzan Prize Foundation, November 24, 2006, http://www.balzan.org/en/prizewinners/paolo-de-bernardis-e-andrew-lange/a--lange-rome--24-11-2006.

18 P. A. R. Ade et al., "Detection of *B*-Mode Polarization at Degree Angular Scales by BICEP2," *Physical Review Letters* 112, 241101 (2014), http://journals.aps.org/prl/abstract/10.1103/PhysRevLett.112.241101.

Chapter 15
POETRY FOR PHYSICISTS

1 Marina Jones, "The Dark Constellations of the Incas," *Futurism*, August 10, 2014, https://futurism.com/the-dark-constellations-of-the-incas/.

2 "...every saint and sinner in the history of our species lived there—on a mote of dust suspended in a sunbeam." Carl Sagan, *Pale Blue Dot: A Vision of the Human Future in Space* (New York: Random House, 2011).

3 Julien de Wit et al., "A combined transmission spectrum of the Earth-sized exoplanets TRAPPIST-1 b and c," *Nature* 537 (2016): 69–72, https://www.nature.com/articles/nature18641.

4 Adam G. Riess, William H. Press, and Robert P. Kirshner, "Is the Dust Obscuring Supernovae in Distant Galaxies the Same as Dust in the Milky Way?" *Astrophysical Journal* 473, no. 2 (1996): 588, http://adsabs.harvard.edu/abs/1996ApJ...473..588R.

5 See Mario Livio, *Brilliant Blunders: From Darwin to Einstein—Colossal Mistakes by Great Scientists That Changed Our Understanding of Life and the Universe* (New York: Simon and Schuster, 2013), for detailed analysis of the "biggest blunder" claim.

6 B. T. Draine, "Interstellar Dust Grains," *Annual Review of Astronomy and Astrophysics* 41 (2003): 241–89, https://doi.org/10.1146/annurev.astro.41.011802.094840.

7 Michael D. Niemack et al., "BFORE: The B-Mode Foreground Experiment," *Journal of Low Temperature Physics* 184, no. 3–4 (2016): 746–53, http://link.springer.com/article/10.1007%2Fs10909-015-1395-6.

8 Paul Steinhardt, "Big Bang blunder bursts the multiverse bubble," *Nature* 510 (2014): 9, https://doi.org/10.1038/510009a.

9 Cal Alumni Association, UC Berkeley, " 'World's Smartest Billionaire': James Simons Is Cal Alumnus of the Year for 2016," https://alumni.berkeley.edu/california-magazine/spring-2016-war-stories/world-s-smartest-billionaire-james-simons-cal-alumnus.

10 "Simons Collaborations," Simons Foundation website, https://www.simonsfoundation.org/collaborations/.

11 CMB-S4: Next Generation CMB Experiment website, https://cmb-s4.org.

12 "Simons Society of Fellows," Simons Foundation website, https://www.simonsfoundation.org/simons-society-of-fellows/.

13 "About the Simons Observatory," https://simonsobservatory.org.

14 John Updike, "Cosmic Gall," available at: https://www.nobelprize.org/nobel_prizes/physics/laureates/1995/illpres/cosmic-call.html.

Chapter 16
RESTORING ALFRED'S VISION

1 Kenne Fant, *Alfred Nobel: A Biography* (New York: Arcade, 1993).

2 "To my nephews, Hjalmar and Ludvig Nobel, I bequeath the sum of Two

Hundred Thousand Crowns each; To my nephew Emanuel Nobel, the sum of Three Hundred Thousand, and to my niece Mina Nobel, One Hundred Thousand Crowns; To my brother Robert Nobel's daughters, Ingeborg and Tyra, the sum of One Hundred Thousand Crowns each." Available at: http://www.nobelprize.org/alfred_nobel/will/will-full.html.

3 Peter Conrad Mayer, personal communication, 2011.

4 Jack Grove, "'Bias' blamed for dearth of female prizewinners," *Times Higher Education*, September 28, 2017.

5 Elaine Howard Ecklund, "A Gendered Approach to Science Ethics for US and UK Physicists," *Science and Engineering Ethics* 23, no. 1 (2017): 183–201, https://link.springer.com/article/10.1007/s11948-016-9751-8.

6 Zuckerman, "Nobel Laureates in Science: Patterns of Productivity, Collaboration, and Authorship," https://doi.org/10.2307/2091086.

7 Ruth H. Howes and Caroline L. Herzenberg, *After the War: US Women in Physics* (Williston, VT: Morgan and Claypool, 2015).

8 Jeremy Venook, "The Political Slant of the Nobel Prize in Economics," *Atlantic*, October 9, 2016, https://www.theatlantic.com/business/archive/2016/10/nobel-factor-offer-soderberg/503186/.

9 Burton Feldman, *The Nobel Prize: A History of Genius, Controversy, and Prestige* (New York: Arcade, 2012).

10 Crawford, *The Beginnings of the Nobel Institution.*

11 "Nobel Descendant Slams Economics Prize," *The Local*, September 28, 2005, https://www.thelocal.se/20050928/2173.

12 Jesse Emspak, "Are the Nobel Prizes Mising Female Scientists?" *Scientific American* LiveScience, October 7, 2016, https://www.scientificamerican.com/article/are-the-nobel-prizes-missing-female-scientists/.

13 Al-Khalili, "Why the Nobel prizes need a shakeup," https://www.theguardian.com/commentisfree/2012/oct/08/nobel-prizes-need-shakeup.

14 Bell Burnell recently won the President's Medal of the Institute of Physics. See "Professor Dame Jocelyn Bell Burnell," President's Medal recipients, Institute of Physics website, http://www.iop.org/about/awards/president/medallists/page_69815.html.

Epilogue
AN ETHICAL WILL

1 Barack Obama, *A Letter to My Daughters* (New York: Knopf, 2010); "Legacy Letters: The Ethical Will," Life Legacies, http://life-legacies.com/index.html; Matthew 5:43–48.

2 Talmud Tractate Pirkei Avot 1:1.

3 Rachel Siegel, "Einstein scribbled his theory of happiness in place of a tip. It just sold for more than $1 million," *Washington Post*, October 24, 2017, https://www.washingtonpost.com/news/worldviews/wp/2017/10/24/einstein-scribbled-his-theory-of-happiness-in-place-of-a-tip-it-just-sold-for-more-than-1-million/?utm_term=.94c4a64e9ec6.

INDEX

Page numbers followed by *f* indicate figures or photographs.